Medicinal Plants of Costa Rica

Medicinal Plants
of Costa Rica

Ed Bernhardt

A Zona Tropical Publication

Disclaimer: Distribuidores de la Zona Tropical, S.A. makes no warranties or representations of any kind concerning the accuracy or suitability of the information contained in this book for any purpose. In no event shall Distribuidores de la Zona Tropical, S.A., its employees, or agents be liable for any direct, indirect, or consequential damages resulting from the use of this book. This publication is primarily for informational purposes and the publisher advises against relying on it in lieu of consultation with a physician or naturopath. Always consult with a physician before taking any form of medication.

A very special thanks is due Agustin Contreras Arias, who contributed a number of beautiful photographs to this book and who kindly took the time to thoroughly read the manuscript before publication.

Text copyright © 2008 Ed Bernhardt
Photographs copyright © 2008 Ed Bernhardt
Additional photographs are by Agustin Contreras Arias and David Katz (www.photo.net/photos/dkatz02) and are reproduced by permission.

The following photographs were contributed by Agustin Contreras Arias: Arnica (p. 13), Borage (p. 20), Dandelion (p. 36), Echinacea (p. 38), Feverfew (p. 42), Greater plantain (p. 48), Mint (p. 64), Passion flower (p. 72), Pisabed (p. 79), Red targua (p. 84), Sarsaparilla (p. 93), Scorpion's tail (p. 94), Stinging nettle (p. 99), Thyme (p. 101), Vervain (p. 108), Winged-leaved quassia (p. 112), Wormseed (p. 113), and Yarrow (p. 114).

The following photograph was contributed by David Katz:
Zornia (p. 118).

The following photograph was contributed by Turid Forsyth:
Pink shower tree (p. 75).

All rights reserved

ISBN: 978-0-9705678-9-5

Printed in China

10 9 8 7 6 5 4 3 2 1

Copy editor: Suzanna Starcevic
Book design: Zona Creativa, S.A.
Designer: Gabriela Wattson

Published by Distribuidores de la Zona Tropical, S.A.
www.zonatropical.net

Table of Contents

Introduction   7

Medicinal Plants   9

Making Herbal Preparations   119

Matera Medica   123

Glossary   125

Bibliography   129

Visual Index   131

Index   137

Introduction

Work on this book really started around 1980, when I first came to Costa Rica in search of adventures in the rainforest. I was a recent graduate of naturopathy and a committed botany student, keen on finding my niche in this exotic country. For about five years, I worked alongside a Costa Rican, Don Carlos Gamboa, who taught me how the indigenous people used medicinal plants. I spent my evenings poring over books on tropical medicinal plants. Sure enough, Don Carlos and his folklore knowledge matched up with the references in the books. I realized how fortunate I was to have this man—a walking encyclopedia on medicinal herbs—as my teacher.

My next step was to put all this useful knowledge into practice. I began to use local medicinal plants to treat my family and neighbors—and myself. I continued to follow current research, and over the next few years I eliminated plants that scientific studies had demonstrated were toxic. During this time, Dr. Luis Poveda of the National Herbarium kindly took me under his wing, sharing the latest botanical information and facilitating access to the vast collection of Costa Rican plants in the National Herbarium. This was particularly helpful for my study of the *Tabebuia* trees (see p. 76), which were an important source of botanical medicines used by the Maya and Inca cultures.

By the early 1990s, I had successfully integrated into my health-care work the use of both natural holistic health-care practices and medicinal plants. Tropical botanicals are powerful healing allies, especially when combined with nutritious food, body work, and mind–spirit healing. For the past 15 years, people have come from around the world to study with me in Costa Rica; this book grew out of my classes on tropical medicinal plants and holistic health therapies.

This is not an exhaustive study of tropical medicinal plants; it is a presentation of tropical plants that are both safe to use and either easy to find in local markets or easy to grow in home gardens (plant descriptions include useful tips on how to grow these herbs).

Often, if you identify symptoms early on, you can treat the illness with herbs and health-care therapies, without relying on expensive medical intervention. And in many cases, medicinal plants and a healthier lifestyle can help you overcome chronic illnesses.

Some plants are included in the text to caution the reader about the toxic compounds researchers have discovered in those species. While those botanicals have been used in the past for medicinal purposes, it is important to select the safe alternatives listed in the book. Medicinal plants are invaluable to your health, but don't forget that this book is meant to complement—not substitute—the medical advice of your doctor.

The plants in this book are arranged alphabetically according to the English common name. Each medicinal plant description includes the English common name—at the top of the page—and the Spanish common name, the Latin species name, and the family name (all appearing in the box next to the photograph). If you are looking for medicinal plants that treat a specific health problem, use the Matera Medica cross-reference guide on p. 123; this index will indicate a group of plants that are reputed to prevent, cure, or alleviate a specific malady. If you know what a plant looks like, but you don't know it's name, the visual index on p. 131 will help you find it. The index on p. 137 lists the plants according to English and Spanish common names and Latin species names.

May you live a healthy life.

Ed Bernhardt, N.D.

Allspice

Geo-distribution:
Native to the East Indies, allspice was brought to the West Indies by the British spice traders. Jamaicans, in turn, brought the plant to Costa Rica at the turn of the last century. It grows throughout Central America and Mexico.

Botanical Description:
This tree grows 8–20 m tall and has a straight trunk and erect branches. The leaves are opposite, oblong to elliptic, and measure 9–20 cm long, 3–9 cm wide. The flowers are white and fragrant; the small fruit is aromatic and turns dark brown when ripe.

Medicinal Uses:
Although used primarily as a seasoning, allspice has also been used as a stomach tonic, an appetite suppressant, and to lower cholesterol. Note: Ingesting the tea should never be used as a justification for abandoning healthy eating habits.

Culinary, stomachic, carminative, digestive.

Preparation:
The leaves can be combined with other tropical herbs to make a delicious tea.

Notes:
The flavor of allspice is said to be a combination of all spices, hence the name. This tree can be planted in the home garden to supply the dinner table with a unique culinary seasoning.

JAMAICA
Pimenta dioica
MYRTACEAE

Aloe vera

Geo-distribution:
Aloe vera is native to South Africa and now grows in most parts of the neotropics. In Costa Rica, it is grown commercially and is popular in patio herb gardens.

Botanical Description:
This is a fleshy plant with small spines along the borders of the smooth, light-green leaves. It grows up to 60–70 cm tall with a yellow spike of flowers.

Medicinal Uses:
Perhaps one of the most important medicinal plants, aloe vera has often been called the "potted physician," since it is used to treat a wide range of health problems. The gelatin inside the leaf is used to alleviate skin problems such as burns, sunburn, insect bites, cuts, bruises, eczema, psoriasis, fungal infections, and acne. It is also used on sprains and strains and as a hair conditioner and beauty treatment. Internally, the gel is used to treat a wide range of digestive problems, including acidosis, heartburn, gastritis, ulcers, constipation, and colitis—along with several liver, gallbladder, kidney, and bladder problems (Gage 1988).

According to the *Journal of the American Pediatric Medical Association*, some studies have found that aloe vera helps to prevent arthritis and to reduce inflammation in affected joints. Aloe can also inhibit the autoimmune reaction associated with certain forms of arthritis, in which the body attacks its own tissues. Researchers writing in *Cancer Immunology and Immunotherapy* found that a compound from aloe (lectin), when injected directly into tumors, activated the immune system to attack the cancer. Killer T cells (white blood cells that bind to invading cells and destroy them) began to attack the tumor cells injected with lectin. Aloe's curative (and preventive) proper-

SÁBILA
Aloe vera
LILIACEAE

ties were also indicated in a study of 673 lung cancer patients in Okinawa, Japan, published in the *Japanese Journal of Cancer Research* (now called *Cancer Science*). "The results of plant epidemiology suggest that aloe prevents human pulmonary carcinogenesis [lung cancer]," stated the researchers. Further, aloe is "widely preventive or suppressive against various human cancers." According to the *Journal of Advancement in Medicine*, aloe vera juice also proved to be an effective part of a nutritional support program for HIV positive patients (Aloe Vera Studies Organization, 1999).

Emollient, digestive, purgative, vulnerary, antiarthritic, antiseptic, antibacterial, antiviral.

Preparation:

The simplest way to prepare aloe is to cut a section of the leaf, remove the spines, and peel it in half. These "patches" can be used to treat skin problems and to treat hair; they can also be used to moisturize the skin (prior to shaving) and to lubricate the skin before a massage. To extract gelatin from the plant for internal use:

1. "Fillet" or slice the fresh leaves.
2. Separate the gelatin from the veins of the leaf. These veins contain a bitter, yellow alkaloid called aloin, which, although it is a powerful antiseptic and antibacterial agent, can cause a strong laxative effect in many people.
3. Place the filleted leaves in 1 L of water and blend.
4. Optional: Add a piece of papaya (and blend) to improve the flavor and enhance the healing properties.

Notes:

Aloe is one of the easiest home garden plants to grow in the tropics. Young vegetative offshoots, called *hijos* in Spanish, can be obtained at local nurseries or often from neighbors. Plant these offshoots in pots filled with sandy soil; keep the pots in a shady place; and water just twice a week. Once the offshoots have taken root, replant them in your garden, either in a sunny spot or one with partial shade. In order to supply your home with enough aloe to meet first aid needs, you should probably cultivate at least a dozen plants.

Annatto

ACHIOTE
Bixa orellana
BIXACEAE

Geo-distribution:
Annatto is a native plant found at lower elevations in the tropical Americas. For centuries, indigenous tribes used the plant as a dye, condiment, and medicine. Today it is cultivated in Costa Rica as a popular food coloring for rice and vegetables.

Botanical Description:
Annatto is a handsome bush that reaches 5 m in height and has pink flowers and ovate, heartlike leaves. The unusual seed capsules are covered with soft red spines and contain numerous seeds coated with a bright reddish-orange resin.

Medicinal Uses:
Although better known as a paste for dye and food coloring, annatto is also used medicinally—it is high in beta-carotene. Tea made from the leaves is reported to be effective in treatment of dandruff, hair loss, and headaches. The seeds and roots have been used to treat fever and diarrhea. A tea made from the seeds and roots is also reported to have a tonic effect on the kidneys and sex glands—some say it is an aphrodisiac.

Hepatic, vulnerary.

Preparation:
To make dye or food coloring, first boil the seeds in water to extract the resin; strain out the seeds; continue boiling down the liquid until it turns into paste.

To make tea that treats dandruff, hair loss, and headaches, boil a handful of leaves in 1 L of water. To prepare a tea that helps relieve fever and diarrhea, select an annatto root; wash it; chop the root into small pieces and boil it in 1 L of water.

Recommended tea serving: 1 cup per day.

Notes:
The seeds germinate readily in pots or plastic nursery bags filled with moist, moderately fertile soil. After several months, the hardy seedlings can be transplanted to permanent sites in the garden. Annatto also serves as an eye-catching ornamental or fence border.

Arnica

Geo-distribution:
Native to the neotropics, arnica grows from Mexico to Argentina. It occurs from low- to midelevations in open fields, along roadsides, and in patios, where it is cultivated as a medicinal herb.

Botanical Description:
Arnica is an herbaceous plant, 10–30 cm tall, with leaves that look much like those of a dandelion (*Taraxacum officinale*, p. 36). Although the two are sometimes confused, arnica leaves are light green above and white below, and slightly dentate. The purple arnica flower is borne on a thin stem; when mature, it produces a seed puff ball that resembles dandelion.

Medicinal Uses:
Arnica works to help treat sore muscles, as well as strains and sprains.

Preparation:
You can make a tincture from the fresh leaves and flowers: Soak one handful of chopped arnica leaves in 250 mL of vodka (or clear rum) and store the concoction in a glass jar with a tight seal; one week before using, shake the mixture once daily. Apply to affected areas for quick relief from pain and stiffness.

Anesthetic, anodyne.

Caution:
This herb is poisonous if taken internally. For topical use only.

Notes:
To propagate arnica, transplant wild plants to your garden.

ÁRNICA FALSA
Chaptalia nutans
ASTERACEAE

Artemisia

Geo-distribution:
Early settlers carried this European herb to the neotropics, where it now grows in most countries of the region, at higher elevations.

Botanical Description:
Artemisia is an erect, glabrous herb, 30-45 cm in height with light-green to grayish leaves. The leaf edges are intensely pinnate.

Medicinal Uses:
Used in emergency situations to treat intestinal parasites, fever, and menstrual problems.

Antiparasitic, febrifuge, emmenagogue.

Preparation:
To make tea, boil 2 or 3 leaflets in 1 L of water. Adults should drink no more than 2 cups per day, children less than 3 tablespoons.

Caution:
This slightly toxic herb may have secondary effects if overused. Nausea, low blood pressure, and vomiting have been reported. Pregnant women should avoid using it.

Notes:
Some gardeners spray this herbal tea on plants to protect their leaves from insects. Artemisia, or mugwort, is easy to grow and can be started from seed or stem cuttings. You can often obtain fresh stem cuttings from Costa Rican herb stands.

AJENJO
Artemisia spp.
ASTERACEAE

Avocado

Geo-distribution:
Avocado trees are native to the neotropics; they can be found at most elevations in Costa Rica.

Botanical Description:
Avocado is a 15–20 m-tall fruit-bearing tree with alternate, elliptic, acutely shaped leaves that are acuminate or rounded at the apex. The yellow flowers have a green, hairy calyx 4.5–6 mm long. The fruit, which can reach a length of 20 cm, contains oily flesh and a hard stony pit.

Medicinal Uses:
In the tropics, the avocado fruit is best known as a food source; however, a few folk cultures mention some interesting medicinal uses. These include boiling the older, dry leaves to treat colds and flu, coughs, fevers, diarrhea, jaundice, stomach problems, and high blood pressure. A maceration of the fresh leaves, used externally, is recommended to treat skin disorders, headaches, and joint pain caused by arthritis or rheumatism.

Expectorant, febrifuge, stomachic, antidiarrheal, abortive, emmenagogue, vermifuge.

Preparation:
To prepare tea, steep old, dry leaves in hot water (3 leaves per cup of water); drink three times per day.

A recipe to tackle parasites, particularly tapeworms, calls for the following:

1. Mix 1 grated seed with 1 L of water.
2. Mix ¼ teaspoon of the solution into a glass of water; drink one serving per day for seven days.

AGUACATE
Persea americana
LAURACEAE

3. Stop for four days.
4. Repeat steps 2 and 3 for a total of four weeks.

The grated fresh seed is also employed as a tincture, applied externally, for treating arthritic pain and skin infections:

1. Mix 1 grated seed into 500 mL of vodka or clear rum and store the concoction in a glass container with a tight seal.
2. Before applying the tincture, shake the concoction vigorously once a day for a total of seven days.

Caution:
An infusion of the young leaves can cause abnormal heart palpitations in some people. Because these young leaves are potentially toxic, you should never take them internally (Vargas Chinchilla 1990).

Basil

Geo-distribution:
Early immigrants from Europe introduced basil to the New World, and it is now found at most elevations in the neotropics. It is a common sight in Costa Rican yards, where it is grown as a medicinal and culinary herb.

Botanical Description:
This annual herb grows 60 cm tall, bearing opposite, ovate, entire (or toothed) aromatic leaves and square stems. The two-lipped flowers vary in color from white to red. More than 20 varieties of this plant exist; sweet basil, holy basil, lemon basil, mint basil, and purple basil are among the most common.

Medicinal Uses:
Basil is a tasty culinary herb with some medicinal properties. If used daily in salads and other dishes, it improves digestion and prevents fermentation and gas in the digestive tract. It is also used as a mild nerve tonic. An infusion of basil tea with honey is useful for treating stomach cramps in babies.

Stomachic, relaxant.

Preparation:
To make a tea infusion, steep a handful of fresh leaves in 1 L of hot, but not boiling, water for 5–10 minutes.

Notes:
It is easy to grow basil in home gardens, from either stem cuttings or seeds.

ALBAHACA
Ocimum basilicum L.
LAMIACEAE

Bitter gourd

Geo-distribution:
A plant widely distributed throughout the Old World and New World tropics. In Costa Rica it grows best along both coasts, though it also extends from the coastal lowlands up to midelevations. It is usually found along roads and in pastures and empty lots.

Botanical Description:
Bitter gourd is a wild, vining plant of the cucumber, or cucurbit, family. It has leaves that are 5-lobed and palmate in shape; yellow flowers with five petals; and a yellow, cucumberlike fruit covered with soft spines and wrinkled skin. The seeds are red with a gelatinous cover that many birds eat.

Medicinal Uses:
Bitter gourd is a popular folk remedy. In Costa Rica, a tea made from the leaves has been used externally for treating skin infections. But more commonly, people drink an infusion of the leaves to alleviate problems with the liver and the pancreas (e.g., hypoglycemia and diabetes) or to stimulate menstruation. The infusion has also been used (as a douche) to treat leucorrhea. The fruit is known as a purgative and as a deworming agent.

Vulnerary.

Preparation:
For external use, steep a handful of leaves in 1 L of water. Strain out the leaves from the infusion and apply the liquid to the skin.

Caution:
Bitter gourd should be used for external application only. During the 1980s, a team of international ethnobotanists who studied folk uses of medicinal plants in the Caribbean region discovered that bitter gourd contains several highly toxic alkaloids that could cause damage to the liver. The team suspected that the high rate of liver disorders in Jamaican children may have been due to their mothers' use of bitter gourd as a medication (Robineau 2005).

Notes:
Although most plants used in folk remedies are safe to use, modern research in plant biochemistry continues to identify plants that may be potentially harmful to humans.

PEPINILLO or SOROSÍ
Momordica charantia
CUCURBITACEAE

Bitterwood

Geo-distribution:
The bitterwood tree is native to the neotropics. In Mesoamerica it generally grows at lower elevations, although in Costa Rica it is found at both low- and midelevations, in secondary and primary forests.

Botanical Description:
Bitterwood is a small tree with unusual leaf bracts on the stems. Young leaves are reddish green with red, 3 cm-long flowers in racemes. The racemes each bear five 1–1.5 cm-long drupes.

Medicinal Uses:
Bitterwood's important role as a folk medicine in Central America dates back to distant history, when it was used by indigenous peoples. The bark and wood of the branches are used to prepare a stomach tonic; they are also used in emergency treatment of fever, diarrhea, and parasitic infections. The tea made from this plant is one of the bitterest in the world!

Stomachic, febrifuge, anthelmintic, antiparasitic, antidiarrheal.

Preparation:
Chop a handful of small branch sections into chips and boil in 1 L of water for 10 minutes. Sip a single cup of tea throughout the day.

To make a tincture, soak a handful of chips in 1 L of vodka. Shake daily for a week before using. Take 3 tablespoons orally per day.

Notes:
Bitterwood tea is also used as a spray that protects plants against disease and that acts as a mild insecticide.

HOMBRE GRANDE
Quassia amara
SIMAROUBACEAE

Bluebush

Geo-distribution:
Bluebush is found throughout Central America (most commonly on the Pacific slope) and occurs in most parts of Costa Rica. It is a common patio plant.

Botanical Description:
This is a bushlike plant that has semi-wooden stems and grows to 1-2 m tall. The leaves are usually light green and lanceolate with opposite veins that are distributed evenly from a main vein. The inconspicuous brownish red flowers are borne on terminal spikes.

Medicinal Uses:
According to folklore about medicinal plant use in Costa Rica, bluebush is used mainly to stem or prevent hair loss. This plant also produces a remarkable blue dye that was used extensively by indigenous tribes. When heated, the leaves produce a rinse that fortifies the scalp and hair. This rinse is particularly suited for those with dark or black hair, since it creates a dark bluish-black sheen.

Preparation:
To prepare a hair rinse, first dry a handful of leaves over a stove or in a skillet, and then boil in about 1 L of water. (Before the age of bleach, bluebush was used as a bluing or brightening agent for clothes. Housewives soaked their white clothes in a diluted solution and then dried them in the sun.)

Caution:
This plant is toxic and should only be used externally, as a hair rinse.

AZUL DE MATA
Justicia tinctoria
ACANTHACEAE

Notes:
Bluebush is easily grown in the home garden from stem cuttings. Mature 30 cm stem sections cut from the base of the mother plant can be planted directly in permanent sites. For better results, however, plant stem sections in plastic nursery bags with moderately fertile soil; keep well-watered in a shady place until the stem sections take root and bear new foliage.

Generally, bluebush thrives in a wide range of soils as long as there is good drainage. Mature plants prefer full sun; no fertilizer is necessary.

Borage

Geo-distribution:
Early European settlers introduced borage to the New World, where it now thrives at higher elevations. The plant is popular as a medicinal herb and is often grown in patio herb gardens.

Botanical Description:
The plant grows to 60–80 cm tall. Its hollow stems bear hairy oval (or oblong-lanceolate) leaves and blue or purplish-to-white star-shaped flowers.

Medicinal Uses:
This herb is used most frequently to reduce fevers and otherwise accelerate the recovery of patients. It is also used to treat colds and flu. Many herbalists prefer to use borage only as an emergency herb, since some people experience secondary effects (see the Caution).

Febrifuge, stomachic, expectorant, alterative.

Preparation:
To make an infusion, steep a handful of fresh leaves in 1 L of water or steep 1 teaspoon of dried herb in 1 teacup of hot water. Drink no more than 2–3 cups per day for up to a week.

Caution:
Some people experience nausea and vomiting with excessive use.

Notes:
Borage thrives at higher elevations, where it is common in home gardens and easy to find in markets and herb stands.

BORRAJA
Borago officinalis
BORAGINACEAE

Broom weed

ESCOBILLA
Sida rhombifolia L.
MALVACEAE

Geo-distribution:
Broom weed, or *escobilla* in Spanish, is one of the most common herbs in the neotropics. It grows in pastures, along roadsides, and in backyards.

Botanical Description:
Broomweed is a shrublike herb with woody stems that grows up to 1 m tall. Its leaves are opposite, serrate, with pointed tips and a reddish tint near the stems. The solitary, yellow flowers, which form in the leaf axils, close at dusk and open again in the morning.

Medicinal Uses:
In the past, this plant was used to make brooms (hence the name), but it was also regarded as an effective medicine for several illnesses. The tea is reported to be a good skin wash for infections or other injuries. Drinking the tea is said to help alleviate colds and flu, coughs, fever, and burning urine.

Expectorant, febrifuge, antiherpetic, demulcent.

Preparation:
To make tea, boil 1 cup of fresh leaves in 3 cups of water for five minutes. Drink one cup of tea between meals.

Carpenter's bush

Geo-distribution:
Carpenter's bush, called *tilo* or *tila* in Spanish, is native to the neotropics. In Costa Rica, the plant grows at lower elevations in fields and near homes.

Botanical Description:
This small, creeping, herbaceous perennial plant has light-green, opposite, ovate-lanceolate leaves. Small, pastel purple flowers grow from the apex of the stems.

Medicinal Uses:
In many parts of the tropics, people use carpenter's bush as an important aid in treating pulmonary infections and as a mild sedative. The name *tilo* is actually erroneous, since in Spanish *tilo* refers to the linden tree (*Tilia* spp.). The name became associated with carpenter's bush because it has the same sedative properties as the linden tree.

Relaxant, nervine, expectorant, pectoral.

Preparation:
To make an infusion, steep a handful of fresh leaves or 1 tablespoon of dried herb in 1 L of water. Suggested dosage: 1-2 cups per day.

Notes:
Carpenter's bush thrives in the tropical herb garden. Look for fresh cuttings in Costa Rican markets and herbs stands.

TILO or TILA
Justicia pectoralis
ACANTHACEAE

Castor bean

Geo-distribution:
The castor bean is native to the Old World tropics and now grows in tropical and temperate climates around the world. In Costa Rica, it occurs at most elevations, usually growing wild in fields, pastures, and at roadsides. You will sometimes see it in rural gardens or nurseries, where it is planted to provide shade for other plants.

Botanical Description:
The castor bean is a perennial shrub or small tree that grows to 1–5 m tall. The palmate, deeply lobed leaves are 10–60 cm wide. The purple to reddish-brown flowers are borne on spikes. These spikes turn into soft, spiny seed capsules with seeds that are entirely black or spotted with brown and gray.

Medicinal Uses:
Castor oil is famous as a laxative, but it has other applications as well. The plant and its seeds have been used in folk remedies as an anti-inflammatory, analgesic, and febrifuge. The oil enhances natural herbal treatments for parasitic worms; it is used as a massage oil for babies suffering from constipation; and nursing mothers use it to heal breast sores and increase milk production. Despite the castor bean's popularity as a medicinal plant, the commercially processed oil, sold in most pharmacies, is much safer to use.

Laxative, anti-inflammatory, analgesic, febrifuge.

Preparation:
Take 1–2 tablespoons of commercial castor oil with a glass of warm water to eliminate constipation and help cleanse the colon of obstructions and putrid waste.

HIGUERILLA
Ricinus communis
EUPHORBIACEAE

Caution:
The fresh castor seeds contain a poisonous compound called ricin, which is rendered non-toxic by heating. Castor oil made commercially has been heat-treated.

Notes:
The castor bean plant, with its distinctive umbrellalike leaves, is used as an ornamental in many Costa Rican home gardens. Costa Ricans believe the plant will keep away moles, which are supposedly repelled by the toxic roots. Plant seeds about 1 inch below the ground.

Cayenne

Geo-distribution:
Cayenne, or hot chili pepper, is native to the neotropics and grows throughout most of the tropical Americas. In Costa Rica, cayenne plants are most commonly seen in private patios.

Botanical Description:
This perennial shrub is 1–3 m tall and bears ovate, obtuse to acuminate, glabrous leaves that are up to 9 cm long. The erect flowers are solitary or in pairs with a corolla 1 cm in diameter; when they are open, the flower petals have ovate, acute lobes. The conical red or orange fruits grow to 10 cm in length.

Medicinal Uses:
Most people think of cayenne as a condiment, but this plant also has some interesting medicinal properties. Its fiery aromatic oil is a strong antiseptic and antibacterial agent that is used to heal cuts and skin infections. The oil also helps to keep the digestive tract free of unwanted microbes, increase circulation and elimination, and provide relief for colds and flu.

Antiseptic, antibacterial, digestive, expectorant, diaphoretic.

Preparation:
Using cayenne to season salad dressings and cooked meals is a good way to prevent health problems. The following "morning liver flush" is a particularly salutary recipe: Mix warm water, lemon, honey, and hot cayenne powder (some folks also like to add crushed garlic or ginger). Drink this concoction in the morning to relieve colds and flu and to help wean yourself off alcohol, tobacco, or coffee. Cayenne powder in capsules helps treat digestive tract problems, heart and circulatory problems, and kidney malfunction. Take 3 capsules per day, along with plenty of water.

Notes:
Dry the seeds before planting them. Cayenne shrubs can last for years, producing spice for your dinner table and an ingredient for home remedies.

CHILE PICANTE
Capsicum frutesens
SOLANACEAE

Cecropia tree

Geo-distribution:
The cecropia tree, native to the neotropics, grows in fields, secondary forests, disturbed areas of primary rainforests, and along roadsides; it is one of the most common pioneer species in the tropics. Costa Rica is home to four mainland species of cecropia—found at various elevations—and all are very similar in appearance: *C. peltata* (usually found on the Pacific slope), *C. insignis* (in wet, humid tropical lowlands), *C. obtusifolia* (in lowland secondary forests), and *C. angustifolia* (in wet tropical highlands). A fifth species, *C. pittieri*, is endemic to Cocos Island.

Botanical Description:
The cecropia is a tall tree—it grows to 20 m—with a narrow, segmented trunk. The leaves are deeply lobed and palmate with a rough, almost abrasive underside. The white or pinkish flowers are borne on clusters of succulent spikes, which some say look like little trumpets; hence the alternate name, trumpet tree.

Medicinal Uses:
Traditionally, an infusion of the leaves was used to help relieve asthma and act as a decongestant for colds, flu, and sore throats. The boiled leaves have been used as a sponge bath for reducing fevers, swellings, and rheumatism.

Pectoral, febrifuge, anti-inflammatory, expectorant, decongestant.

Preparation:
To make an infusion, steep 2 leaves in 1 L of hot water for 15 minutes. Drink 2 cups of tea per day for no more than 3 days; you

GUARUMO
Cecropia spp.
CECROPIACEAE

can also use the liquid as a gargle for sore throats. Boil several large leaves in 4 L of water to make a sponge bath.

Caution:
Although the cecropia tree has been used extensively as a folk remedy, some studies have shown the leaves to have a toxic effect on mice at a low internal dosage

(Arvigo and Balick 1993). For this reason, caution should be used when administering cecropia internally (do not drink the infusion for more than 3 consecutive days).

Notes:
Several ant species colonize the inside of the trunk in a symbiotic relationship with the cecropia tree. The ants provide protection from invaders and, in return, the tree provides food in the form of a sap that is exuded from the base of the leaf stems. Because of its large size, the cecropia tree is not recommended for home gardens. Look for the tree in the wild—it is easy to find.

Chamomile

Geo-distribution:
Early settlers brought these European herbs to the neotropics, where today they generally occur at higher elevations.

Botanical Description:
Two types of chamomile grow in the tropics—German chamomile (*Matricaria chamomilla*) and Roman chamomile (*Anthemis noblilis*). They are difficult to distinguish, but German chamomile reputedly has greater medicinal value. It has round, downy, hollow, furrowed stems that may be procumbent or rise upright to a height of 35 cm. The leaves are pale green, bipinnate, sharply incised, and sessile. The daisylike flower heads are yellow and white.

Medicinal Uses:
Chamomile makes perhaps one of the most popular herbal teas. It is known to improve digestion and to act as a mild calmative; many people take it to induce sound sleep. It is also an effective eyewash for sties and sore eyes. And, chamomile can also be used to make a rinse for light or blond hair.

Digestive stimulant and tonic, carminative, relaxant, diaphoretic.

Preparation:
To create an infusion, steep a teaspoon of dried chamomile in one cup of water. Drink 1–2 cups per day.

Caution:
Although chamomile makes a pleasant, healthy tea, drinking it in excess can cause the stomach to release too much acid and may result in indigestion.

Notes:
Chamomile grows in many home gardens and it is readily available in markets and herb stands. If you want to grow chamomile in your own garden, dry the seeds from the flowers and plant them in flats or directly in garden beds.

MANZANILLA
Matricaria chamomilla
and *Anthemis noblilis*
ASTERACEAE

Chan

Geo-distribution:
Chan is native to Central America; in Costa Rica, it usually occurs at low elevations on the Pacific slope. This plant grows in open fields and pastures, where it is frequently considered an invasive plant. It also grows along roadsides and in vacant lots. The Costa Rican name is derived from the indigenous Térraba word for the plant—*tšián-ko*.

Botanical Description:
Chan is an annual herbaceous plant that grows to 1.5 m tall. The aromatic, opposite leaves are pubescent, crenate, and dentate. Flowers form on dense spikes. Chan has seed capsules that contain black, triangular seeds.

CHAN
Hyptis suaveolens
LABIATAE

Medicinal Uses:
Traditionally, chan seeds have been used to make a mucilaginous beverage that relieves indigestion, gastritis, and constipation. Chan leaves are prepared as an infusion that has been reported to help reduce high blood pressure.

Nutritive, digestive tonic, laxative.

Preparation:
To make *fresco de chan*, a common drink in Costa Rica, soak 1 tablespoon of chan seed in 2 L of water overnight. (The following day, the seeds—suspended in the thick liquid—take on the appearance of frog eggs, with a mucilaginous exterior and a black center.) Add honey or sugar to taste. Although consumption of this drink has declined in recent years, it is still served in many Costa Rican diners (known as *sodas*). Chan has many of the same healthful properties of flax seed; both contain soluble fiber and mucilaginous compounds that improve digestion.

For a mintlike tea, steep a handful of fresh leaves in 1 L of water for 20 minutes.

Notes:
You can either buy chan seeds at some Costa Rican markets or collect them from wild plants. Plant the seeds in your home herb garden—two or three plants will produce a sufficient number of seeds to supply your home with *fresco de chan*. Since chan plants can grow quite large, it is best to plant them on the edges of the garden. Chan does best in full sun and when planted in well-drained soil. The plant is hardy and has no significant pests or diseases.

28 / Medicinal Plants of Costa Rica

China root

CUCULMECA
Smilax lanceolata
SMILACACEAE

Geo-distribution:
China root principally grows in the primary rainforest habitats of Costa Rica and other Central American countries.

Botanical Description:
China root is a vine that grows in the rainforest, climbing from the forest floor to the canopy. Its thorny stem carries shiny, smooth, lanceolate, alternate leaves that are about 10 cm long and 5 cm wide. The flowers are small, inconspicuous, and green, and the seed capsules are brown. The tuber, which is used to prepare medicine, is reddish.

Medicinal Uses:
More than 200 species of *Smilax* are widely distributed throughout the world's tropical climates. Sarsaparilla, often called "tropical ginseng," is perhaps the best known because of its popularity as a tonic tea.

In Costa Rica and other Central American countries, China root has a reputation as a medicinal ingredient in blood-purifying, sudorific, and revitalizing tonics. China root is also said to treat anemia, general weakness, impotency; skin, kidney, and bladder problems; and it is recommended for cleansing toxins or reducing acidity in the body.

Tonic, nutritive, diuretic, antiherpetic, depurative.

Preparation:
Wash the tuber thoroughly, cut into small pieces, and dry in the sun. Boil about 5 or 6 g per liter of water or milk for at least 20 minutes. In Belize, cinnamon and nutmeg are used to spice up the drink (Arvigo and Balick 1993).

Notes:
China root is only found in the rainforest; it is not grown in the home garden. However, herb stands and health food stores in Costa Rica sell the tuber or products made from this plant.

Cilantro

Geo-distribution:
Cilantro, or coriander, is originally from the Old World; early European settlers introduced the herb to the neotropics. In Costa Rica, it grows successfully at most elevations.

Botanical Description:
Cilantro has a round, finely grooved stem with pinnately decompound leaves. The flowers are white to pink in compound umbels. The seeds are brown with a strong, aromatic scent.

Medicinal Uses:
Cilantro is one of the most nutritious edible herbs; its seeds and leaves have been used for thousands of years as seasoning. Used regularly, cilantro acts as effective preventive medicine, but it can also be used to treat an upset stomach, vomiting, and diarrhea.

Stimulant, aromatic, carminative, nutritive.

Preparation:
To make an infusion, steep 1 tablespoon of seeds or a handful of leaves in 1 L of water. Combine with mints (see Peppermint and Spearmint, p. 73, and Mint, p. 64) to enhance the effect of the cilantro.

Caution:
The seeds can have a slightly narcotic effect if used excessively.

Notes:
Culantro coyote (*Eryngium foetidum* L.), or wild coriander—see photo bottom left—is native to the neotropics. In an example of parallel evolution, this plant independently developed the same aromatic oil as cilantro (from the Old World). Since its scent is identical to that of cilantro, wild coriander is also used as food seasoning. As a medicinal herb, wild coriander is also used to treat stomach problems, vomiting, and diarrhea.

CULANTRO
Coriandrum sativum
UMBELLIFERAE

Cinnamon

CANELA
Cinnamomum verum
LAURACEAE

Geo-distribution:
Cinnamon is native to Sri Lanka, but it is now cultivated and naturalized throughout the neotropics, where it occurs at lower elevations.

Botanical Description:
The cinnamon tree, which has very aromatic bark, grows 8–15 m tall. The leaves are ovate-lanceolate or obtuse (or acute); they grow to 10–18 cm long. The flowers are yellowish white, borne on panicles; the fruit is about 1.5 cm long.

Medicinal Uses:
Cinnamon is commonly known as a spice, but it also has some useful medicinal properties. The powder is used to treat diarrhea, staunch bleeding, and heal skin infections.

Carminative, digestive, antiseptic, antidiarrheal.

Preparation:
When young children have diarrhea this formula comes in handy: Mix 1 tablespoon of cinnamon powder with honey. Children rarely turn down this sweet medicine and it works for most cases of diarrhea.

Notes:
Seedlings are often available in nurseries; they grow well in home gardens at lower elevations.

Citrus

CÍTRICO
Citratus spp.
RUTACEAE

Geo-distribution:
Originally from Asia, citrus fruits are now cultivated throughout the neotropics.

Botanical Description:
The majority of citrus species are small to medium trees with glabrous, ovate leaves—sometimes with spines on the stems—and white, fragrant flowers.

Medicinal Uses:
Citrus fruits are one of nature's greatest gifts to those who want to live a healthy life. Lemons, limes, grapefruits, and oranges all contain vitamin C and bioflavonoids. The leaves and rind also contain essential oils, dominated by limonene. This component is slightly sedative, antispasmodic, and expectorant, hence the inclusion of citrus leaves or peels in many folk remedies for calming nerves and treating colds and flu. Orange peels make a flavorful tea that improves digestion.

Expectorant, relaxant.

Preparation:
Drinking plenty of orange juice when you have a cold or flu can help alleviate the uncomfortable symptoms and accelerate the healing process. For a powerful natural antihistamine that dries up the sinus cavities, blend 2 cups of orange juice with 1 peeled onion and 1 clove of garlic. The sweet juice masks the pungent garlic and onion taste. You will be surprised at how good it tastes and how effective it is in reducing that swollen-head feeling.

The morning "liver flush" is another potent cleansing drink made from lemon, warm water, honey, and hot cayenne powder. Some folks also add crushed garlic or ginger to the recipe. The drink is effective against colds and flu, or if you are trying to wean yourself off alcohol, tobacco, or coffee.

To prepare tea, first hang orange peels in a warm, dry place until they are well dried. Chop the peels into small pieces and store them in an airtight jar to protect them from mold. You can use the dried peels with other herbs—including lemongrass, Jamaican hibiscus, and ginger—to create a delicious tea.

Use young, tender leaves of citrus trees, particularly *naranja agria* (*C. aurantium*), to make a tea that calms the nerves and helps relieve headaches.

To make an infusion, steep 2–3 young leaves per cup of tea. Drink throughout the day and before bed. For headaches, use the tea to make a compress that you apply to the head.

Notes:

Citrus trees are ideal for the home garden; grafted varieties of many species are available in nurseries. Such varieties often begin producing fruit after three years and provide lots of citrus fruits for years to come.

Coconut

Geo-distribution:
This palm is commonly found at lower elevations in the neotropics.

Botanical Description:
The coconut palm grows up to 20 m tall with a trunk that reaches 50 cm in diameter. The fronds are actually pinnate leaves that grow up to 5 m long. The fruit, which contains a large 20-cm-long seed filled with liquid, is ovoid and green. The endosperm is a gelatin when immature (when ripe, it turns into a white, oily fiber).

Medicinal Uses:
The coconut is a fine example of a tasty food that is also used as a medicine; it has been used for a variety of treatments around the world. The young coconut, or *pipa* as it is known in Costa Rica, is an excellent food for infants and children. Both young and ripe coconuts help in treating indigestion, diarrhea, gastritis, ulcers, hepatitis, and low energy.

Coconut water (*agua de pipa*, in Spanish) is considered a diuretic and is often prescribed with pineapple juice or spiral flag for kidney problems. It is recommended in the treatment of parasitic worms and many illnesses, including heart and circulatory infirmities, liver ailments, and problems with digestion.

The oil can be used to treat the skin and hair.

Nutritive, digestive aid, diuretic, circulatory aid, hepatic, antiparasitic.

Preparation:
To prepare a nutritious meal for infants or children, mix young coconut with mashed or blended bananas. Coconut milk extracted from the grated mature coconut meat is a rich, nutritional drink; mix it with other tropical fruits or pour it over cereals.

Notes:
Coconuts are ideal for the tropical garden. The trees are decorative and the fronds and other parts of the plant are used as building materials.

COCO
Cocos nucifera
ARECACEAE

Cornsilk

PELO DE MAÍZ
Zea mays
POACEAE

Geo-distribution:
Corn is native to the neotropics.

Botanical Description:
An annual plant, corn grows up to 2 m in height with long, lanceolate leaves of 1 m and flowering panicles. The seeds form spikelike inflorescences on a woody stem; these are the familiar ears of corn.

Medicinal Uses:
Many people enjoy corn on the cob, but are not aware that corn silk has medicinal properties. In fact, it is one of nature's most effective medicines for kidney problems. Inexpensive yet effective, corn silk has been used to treat retained or burning urine, kidney stones, and bladder infections.

Demulcent, diuretic.

Preparation:
To assure yourself of a ready supply of medicine, take a minute when preparing corn on the cob to select the best corn silk. Place it in a warm, dry place until it is thoroughly dried, then store it in an airtight container to preserve its properties. To make an infusion, put 1 tablespoon of dried corn silk in 1 L of water. Sip up to 3 cups per day.

Notes:
Select organically grown corn to avoid agrochemical residues commonly found in the silk of commercially grown corn.

Dandelion

Geo-distribution:
Dandelion, which also occurs in North America and Europe, grows above 1,000 m in the neotropics.

Botanical Description:
Dandelion is a perennial herb with flowers that rise from the center of the serrate leaves. The flowers are about 15 cm tall with a simple yellow head. The leaves are sharply dentate and dark green.

Medicinal Uses:
This modest little plant, which most people think of as a noxious weed, is actually one of nature's greatest medicines! It has been used in the treatment of a wide variety of illnesses, including gout; inflammation of the bowels; dyspepsia; rheumatism; dropsy; kidney, spleen, pancreas and liver problems; anemia; skin diseases; jaundice; diabetes; eczema; urogenital problems; acne; age spots; water retention; tonsillitis; low blood pressure; boils; insomnia; psoriasis; senility; hypoglycemia; hemorrhage; heartburn; gall stones; fatigue; cancer; bronchitis; constipation; and cramps!

Diuretic, aperient, hepatic, cholagogue, depurative, stomachic, tonic, mild laxative, renal tonic.

Preparation:
Young dandelion leaves can be mixed into salads; the mature root can be made into a tonic that treats the entire body. The root is the most potent part of the plant. To make a decoction, boil 1 teaspoon of the powdered root in 1 cup of water. Drink 1–3 cups daily between meals. An infusion of 40 g (about a handful) of the fresh leaves in 1 L of water is also efficacious. To make a tincture, soak 60 g of dried root in 1 L of vodka; use 1–3 teaspoons per day.

Caution:
It is very important not to confuse dandelion with the very similar arnica (p. 13), which you should never consume internally.

Notes:
Dandelion thrives at higher elevations of the tropics. Grow the plant around your home and harvest throughout the year. Chicory, a close relative of dandelion, has similar properties; the dried root of this plant can be used as a coffee substitute.

DIENTE DE LEÓN
Taraxacum officinale
ASTERACEAE

Dwarf poinciana

HOJA SEN
Caesalpinia pulcherrima
CAESALPINIOIDEAE

Geo-distribution:
The dwarf poinciana, which is native to the neotropics, is planted as a backyard ornamental from sea level to higher elevations.

Botanical Description:
Dwarf poinciana is a shrub or small tree that grows to 5 m tall. The brittle branches carry large leaf stems that have 3–9 pairs of pinnate leaves with 6–12 pairs of leaflets. The flowers are red or bright yellow; the plant produces a fruit pod or capsule that is 12 cm long.

Medicinal Uses:
The leaves are primarily used as a laxative. You can also soak the leaves in hot water for a bath treatment that is said to treat sluggishness and depression.

Laxative, anodyne, analgesic.

Preparation:
Boil a handful of fresh leaves in 1 L of water for five minutes. Take 1–3 cups per day for constipation. Soak several large handfuls of leaves in hot tub water for bathing.

Echinacea

Geo-distribution:
Originally from the mid-western plains of North America, echinacea (or cone flower) is now widely planted at higher elevations in the neotropics.

Botanical Description:
Echinacea is a perennial herb. It bears hairy green lanceolate leaves up to 70 cm long and conical composite flowers of purple-reddish hues.

Medicinal Uses:
Echinacea was originally used by the indigenous tribes of the mid-western plains of North America as an emergency remedy for many illnesses. It is a strong tonic and blood-purifying herb that is used in the treatment of colds, flu, coughs, sinus problems, fevers, infections, immune-response deficiencies, venomous snakebites, arthritis, and rheumatism.

Expectorant, febrifuge, antibacterial, antiviral, tonic, alternative, depurative, antiarthritic.

Preparation:
Many parts of the echinacea plant can be used for medicine. To make an infusion, steep a handful of leaves and/or seed capsules in 1 L of water for 20 minutes. Drink frequently during the day.

The root is the most potent part of the plant. Select from among the more mature plants in your garden and then wash, clean, and thoroughly dry the roots. Store them in an airtight container to prevent them from going moldy or rotting. To make an infusion, steep 1 teaspoon of the powdered root in one cup of water. Use 60 g of powdered root or 180 g of fresh root in 1 L of vodka to make a good tincture (take 3 teaspoons per day).

Notes:
Echinacea thrives in tropical highland gardens; it fares marginally at lower elevations. Plant the seeds, which can be obtained from international seed companies, in germination flats and transplant the seedlings to the garden when they are hardy. This plant grows best in fertile soil with full sunlight.

ECHINACEA
Echinacea purpuria and *E. longata*
ASTERACEAE

Elderberry

SAÚCO
Sambucus mexicana
CAPRIFOLIACEAE

Geo-distribution:
Elderberry grows along streams and marshy areas at higher elevations of tropical Mesoamerica. In Costa Rica, it is also planted in home gardens for medicinal use.

Botanical Description:
Elderberry is a shrub with gray bark, scaly stems, and compound leaves of five ovate-lanceolate leaflets with serrated edges. The small, numerous, pastel yellowish-white flowers are borne on umbels. You will rarely get to see the small black fruits, which are 6 mm in diameter, because birds feed on them as they ripen.

Medicinal Uses:
The flowers have been used as an emergency medicine for fevers, colds, coughs, bronchitis, measles, mumps, and flu. A rinse made from the leaves is said to help stimulate hair growth. The leaves, bark, and root have been used for skin problems.

Febrifuge, diuretic, expectorant, antiherpetic.

Preparation:
You can either use fresh flowers or dried flowers that you have earlier collected and stored. Steep a handful of fresh flowers

(or 1–2 tablespoons of dried flowers) in 1 L of boiled water. Drink 1–3 cups per day. A handful of the leaves, roots, and bark can be boiled in 1 L of water to prepare an external wash for skin conditions.

Caution:
Although the flowers and berries are safe to consume, the leaves, roots, and bark are said to be slightly toxic and are not recommended for internal use (Lust 1974).

Notes:
Elderberry is a useful medicinal plant that is easy to grow in home gardens throughout Costa Rica's higher elevations. Simply take a few 30 cm-long woody stem cuttings from mature shrubs and plant them 15 cm deep in plastic nursery bags with prepared potting soil. Water them and keep them in a shady place until they are well established. Slowly bring them into full light. Finally, plant them in permanent sites around the garden border; they will grow into large bushes, so they need some space. Although many campesinos simply plant stem cuttings directly in the soil in their permanent site, it's best to follow the recommended extra steps in order to ensure that the plants grow successfully.

Eucalyptus

EUCALYPTO
Eucalyptus cinerea
and other species
MYRTACEAE

Geo-distribution:
Eucalyptus was carried from Australia to the neotropics. In Costa Rica, eucalyptus species are commonly found in fields, on tree farms, and in parks and other public spaces.

Botanical Description:
Eucalyptus trees usually grow to 30 m in height; the distinctive bark has a peeling, paperlike surface. The leaves of all eucalyptus varieties contain eucalyptol, the distinctive aromatic oil associated with the tree.

Medicinal Uses:
The leaves of eucalyptus trees are famous for their aromatic oils, which can be useful in treating upper respiratory infections such as coughs, colds and flu, sinus problems, and asthma. Externally, the oil is used to treat skin infections.

Antiseptic, expectorant, stimulant.

Preparation:
To treat respiratory problems, prepare a vapor bath by putting a handful of leaves or a few drops of oil (available in health food stores) in 2 L of water. Place a towel over your head and breathe in the vapors. A bath with a strong infusion of the fresh leaves is used to treat skin problems. Sometimes the fresh leaves can be used to make an infusion as an emergency treatment for fever (but see Caution).

Caution:
Drinking excessive amounts of tea may cause secondary effects such as nausea and vomiting.

Feverfew

Geo-distribution:
Native to Europe, feverfew was introduced to the neotropics, where it grows at higher elevations.

Botanical Description:
This perennial herbaceous plant has daisy-like flowers on stems approximately 50 cm tall. The flowers are yellow with white rays. The green leaves are alternate and petiolate with leaflets inclining to ovate and dentate.

Medicinal Uses:
As its English name implies, feverfew is used to treat fevers. It is also effective as a tonic for the nervous system and has been used to help alleviate colds and flu, arthritis, indigestion, muscle tension, worms, and the effects of alcohol withdrawal. Until recently, women commonly took a feverfew tonic after giving birth. Some people with migraines find relief by eating small quantities of the fresh leaves each day.

Tonic, carminative, bitter, emmenagogue, vermifuge, stimulant.

Preparation:
Feverfew makes a very bitter tea. Steep 1 teaspoon of leaves in a cup of boiling water. Drink 1–2 cups per day.

Notes:
Feverfew thrives in home gardens and it is commonly found in Costa Rican markets.

ALTAMISA
Tanacentum parthenium
ASTERACEAE

Garlic

AJO
Allium sativum
LILIACEAE

Geo-distribution:
Settlers from the Old World introduced garlic to the neotropics.

Botanical Description:
Garlic is a perennial plant with a compound bulb and an erect stem with parallel veins from which grow long, flat, linear leaves. Small garlic bulbs grow below the base of an umbel of white flowers.

Medicinal Uses:
Garlic has been used for centuries as both food and medicine; eating the fresh bulbs is a good way to prevent health problems. It has been proven effective in reducing high blood pressure and cholesterol, and is known to be a strong antibacterial, antiviral, and antifungal agent. Large doses of allyl, a major component of garlic, have been shown to act as a natural antibiotic. Garlic also helps prevent unwanted microbes from colonizing your intestines or cells and makes it harder for worms and other parasites to establish themselves in your digestive tract.

Garlic pills are useful in relieving a broad range of acute or chronic illnesses, including asthma, arthritis, cancer, circulatory problems, colds, flu, infections, insomnia, liver disease, sinusitis, ulcers, and yeast infections (Murray 1991).

Antibacterial, antiviral, antifungal, depurative, hepatic, expectorant, antiparasitic, hemostatic.

Preparation:
Garlic can be prepared in several ways for medicinal purposes. A thin slice of raw garlic can be taped to warts or boils for relief; apply a fresh piece in the morning and evening for 3-7 days. When taking garlic capsules, the recommended dosage is 2 capsules every 2 hours during the day. Use fresh garlic in green salads to strengthen overall health.

Notes:
Garlic plants grow from cloves that you can plant either in the garden or in pots and containers around the home.

Ginger

Geo-distribution:
Originally from Asia, ginger now occurs at low elevations in the neotropics.

Botanical Description:
Ginger is a succulent herbaceous plant that grows from a rhizome that has a pungent taste. Lanceolate leaves, which are 20 cm in length and 2 cm in width, grow on a central stem. These leaves and stems wither in the Costa Rican dry season. The flowers are conelike, yellow or red, and occur near the base of the plant.

Medicinal Uses:
Ginger has been used for centuries as a spice and medicine. In India it is known as a universal medicine. This root is a perfect example of food as preventive medicine, particularly in curries. Ginger oils are antibacterial and antiviral; if used regularly, they can help prevent infections. Ginger, whether hot or cold, combines well with other herbs such as lemongrass, Jamaican hibiscus, and mint. In fact, it goes well with many of the medicinal plants in this book, because it acts as a good "carrier" herb,

JENGIBRE
Zingiber officinale
ZINGIBERACEAE

stimulating absorption and increasing circulation of the herbs it accompanies. More than half of the nearly 2,000 traditional Chinese herbal formulas contain ginger!

Ginger acts as a cleansing and tonic herb for the entire body. Use it to treat colds and flu, sore throats, morning and motion sickness, and circulatory, digestive, kidney, and bladder problems. Applied externally as a bath or massage oil, it alleviates sore muscles, strains, sprains, and arthritic discomfort. Ginger is a natural stimulant and a good substitute for coffee; it is also a tonic herb for the sex glands and is reputed to be an aphrodisiac.

Tonic, expectorant, sudorific, stomachic, analgesic, antibacterial, diuretic.

Preparation:

Ginger is used in many ways. To make a massage oil for strains, sprains, or arthritic pain, mix 2 tablespoons of macerated root or juice extract with 60 g of sesame oil. For a hot chest compress that treats upper respiratory infections and sore muscles, strains, and sprains, grate 2 ginger roots and heat them in 4 L of water. Ginger is also used in a poultice for lower back problems, joint pain, and other related problems; in a blender, combine 30 g of ginger, turmeric, and aloe (remove the spines first).

Sucking on small cross-sections of fresh ginger will often relieve a sore throat.

Ginger is most effective as an infusion. Use 1-2 g of powdered ginger or 5 g of fresh ginger for each cup of tea; preparing the infusion in a teapot will prevent the aromatic oils from evaporating.

Ginger is an ingredient in curry, a tasty and healthy addition to rice and vegetable dishes. To prepare a curry, combine the following powdered herbs: 4 parts ginger, 4 parts turmeric, 2 parts ground cilantro seeds, and 2 parts cumin.

Powdered ginger can be used in empty gelatin capsules to treat digestive problems, morning sickness, and motion sickness. Take 2-6 capsules per day.

Notes:

Ginger thrives in the tropical home garden. In Costa Rica, rhizomes can be planted close to the surface of the soil (leaving the root partially exposed). Plant them in their permanent sites during April and May. Ginger can be harvested each year during the dry season. This plant requires little fuss and provides abundant quantities of seasoning and medicine.

Golden shower tree

Geo-distribution:
The golden shower tree, native to India and Egypt, arrived in the New World tropics during the early colonial period and adapted well to lower elevations. It flourishes in the drier coastal areas of Costa Rica, including Puntarenas, a port town on the Pacific coast. In these areas, the golden shower tree is commonly seen in parks and along avenues.

Botanical Description:
This tropical ornamental tree grows to 10 m in height and carries four to eight pairs of large compound leaves. The tree is esteemed for its beautiful 30–40 cm long racemes of yellow flowers. These flowers produce brown capsules (up to 50 cm long) with numerous cells that contain seeds and a dark pulp.

Medicinal Uses:
The seed capsule is used as a laxative.

Laxative.

Preparation:
To make an infusion, steep a mature seed capsule in 2 L of hot water; drink half a cup of tea before going to bed to help relieve constipation.

Caution:
An overdose can cause a cathartic reaction (diarrhea and intestinal cramping).

Notes:
Collect seeds during Costa Rica's dry season (December–April). Fill plastic nursery bags with good soil and plant the seeds 2-3 cm deep. Keep the seeds well-watered until they germinate; when the seedlings are about 30 cm tall, transplant them to permanent sites. Golden shower trees prefer full sun and well-drained soil.

CAÑAFÍSTULA
Cassia fistula
CAESALPINIOIDEAE

Gotu kola

GOTU KOLA
Centella asiatica (formerly *Hydrocotyle asiatica*)
UMBELLIFERACEAE

Geo-distribution:
Native to Asia, gotu kola now also grows at lower elevations in the neotropics. In Costa Rica, this plant is occasionally seen in commercial farms or home gardens, where it is grown for medicinal use.

Botanical Description:
Gotu kola is a low creeping herbaceous plant that grows on running stems. The 4–5 cm wide leaves have a characteristic horseshoe shape.

Medicinal Uses:
Gotu kola has been used for centuries in India and China to revitalize brain cells and as treatment for nervous disorders, colds and flu, lung problems, urinary tract infections, and herpes. It is a great substitute for coffee and it helps against stress and fatigue. It is one of the best herbal medicines for skin infections. Gotu kola has also been used traditionally to help alleviate rheumatism, blood diseases, mental disorders, high blood pressure, congestive heart failure, sore throat, tonsillitis, hepatitis, psoriasis, eczema, venereal diseases, measles, and insomnia.

Nervine, tonic, expectorant, vulnerary.

Preparation:
To make an infusion, steep 1 teaspoon of dried leaves in 1 cup of hot water for 10–15 minutes. To treat skin problems, prepare a maceration of the fresh leaves.

Notes:
Gotu kola thrives in the tropical herb garden, particularly at lower elevations. The plant prefers locations where soil is damp throughout the year. North American and European companies sell seeds; Costa Rican markets often sell fresh plants suitable for transplanting.

Greater plantain

Geo-distribution:
The greater plantain is a common plant in many parts of the world; in the neotropics, it occurs at higher elevations.

Botanical Description:
This plant has broad ovate leaves and a thick main stem. The flower stalks are 15-40 cm long, with slender spikes of greenish flowers.

Medicinal Uses:
Greater plantain is primarily known as a treatment for skin problems; the fresh leaves are used to make a maceration for this purpose. Fresh or dried leaves are used to treat colds, flu, sore throat, liver problems, and coughs.

Astringent, demulcent, diuretic, expectorant, hepatic.

Preparation:
Mash fresh leaves to make a poultice for skin infections. To make a tea, steep several fresh leaves or 1-2 g of dried leaves in a cup of hot water. Drink 3-6 cups a day until results are obtained.

LLANTÉN
Plantago major
PLANTAGINACEAE

Guapinol

GUAPINOL
Hymenaea courbaril
CAESALPINIOIDEAE

Geo-distribution:
Native to the neotropics, guapinol grows in Mexico, the West Indies, Central America, Peru, and Brazil. The tree occurs in low- and midelevation rainforests and savannahs. Guapinol is best known for its resinous sap, which has been used to make varnish.

Botanical Description:
Guapinol is a large tree—it grows to 15 m tall. The cylindrical trunk is covered in smooth, gray bark. The glossy, dark green alternate leaves grow in pairs that look like an ox hoof. The white flowers are borne on panicles and form large, brown seed capsules (5–15 cm long) that are embedded in a fibrous, edible pulp. The pulp has an unusual odor; on the Caribbean coast of Costa Rica, people call the tree "stinky toe." Guapinol trees are highly esteemed for their hard, red wood; as a result, the species has been overexploited for timber and is now considered endangered.

Medicinal Uses:
Guapinol is one of the most useful trees in the tropics. The nutritious, edible pulp that surrounds the seeds contains high levels of iron, protein, and B vitamins, and it is used to help treat anemia. For centuries, indigenous tribes have used the red sap from the tree trunk as a general tonic, as well as a treatment for upper respiratory problems. Before the advent of modern antibiotics, the sap was used as medicine for tuberculosis in South America. The leaves

and inner bark are used in the treatment of hypoglycemia and diabetes, stomach disorders, and diarrhea.

Astringent, antiseptic, expectorant, depurative, tonic, nutritive, stomachic.

Preparation:

Cook the pulp with milk and honey to make a nutritive drink that is suitable for both children and adults. To prepare a tincture from the sap, follow these instructions:

1. Make an incision in the trunk bark.
2. Collect the reddish amberlike sap with a spoon.
3. Store the sap in a small glass jar.
4. Mix the sap with vodka or clear rum.

This tincture can be used for upper respiratory problems or as a general tonic. Take 1 teaspoon 3 times a day.

To make an infusion, steep a handful of leaves in 1 L of hot water; to make a decoction, boil a handful of inner bark in 1 L of water. Drink 1–3 cups of tea per day.

Notes:

Guapinol is a hardy tree that is ideal for reforestation projects; because of its large size, however, it is inappropriate for small home lots. This species is under serious threat from logging: When extracting sap from the tree, you should try not to damage the tree. After extracting the sap, treat the incision with a sealant to prevent disease from entering the cut.

Guava

GUAYABA
Psidium guajava
MYRTACEAE

Geo-distribution:
Native to the neotropics, guava occurs most commonly at lower elevations, in pastures and along roads.

Botanical Description:
The guava tree grows up to 10 m tall; its trunk has a peeling, dark brown bark. The leathery, opposite leaves are 8–14 cm long and 3–6 cm wide. At the ends of the branches are flowers with white petals. The fruit is yellow or pink when ripe.

Medicinal Uses:
Guava trees are best known in the tropics for their fruit, which is made into delicious preserves, but few people realize that the leaves have potent medicinal properties. A decoction of the leaves is an effective treatment for diarrhea and sore or bleeding gums; it also serves as a douche for leucorrhea.

Astringent, antimicrobial, antidiarrheal.

Preparation:
To prepare tea, boil a handful of leaves in 1 L of water. Drink 3 cups per day. The tea can also be used as a skin wash or douche.

Caution:
Use only as an emergency treatment; the leaves contain tannins and excessive use may result in secondary effects such as nausea or vomiting.

Notes:
Guava is common in the neotropics. It is often easy to find should you need a handful of leaves to prepare tea.

Gumbo limbo

Geo-distribution:
Native from Mexico to Venezuela (from sea level to midelevations), gumbo limbo is used throughout Costa Rica to create living fences that border pasture lands.

Botanical Description:
Gumbo limbo, which grows up to 10 m, has a green trunk and distinctive red bark that peels. The leaves are deciduous compound, with 5–7 leaflets (each 5–12 cm long). This tree bears minute, fragrant flowers that are greenish yellow; its round fruits (6-10 mm long) are tinged with red.

Medicinal Uses:
Some people ingest the bark of this tree as an emergency decoction to treat fever, diarrhea, vomiting, internal parasites, kidney and bladder problems, infections, skin problems, colds and flu, headaches, and sunstroke. Soaking in a bath of the tea is said to treat skin infections and rashes.

Febrifuge, antiparasitic, antidiarrheal, astringent, antiseptic, antimicrobial, expectorant, diuretic.

Preparation:
Costa Rican campesinos use machetes to cut a 10-by-30 cm piece of bark, then chop it and boil it in about 4 L of water for 10 minutes or more. Sip 1 cup of tea throughout the day for up to 6 days. The tea can also be used for a skin wash or a douche.

Caution:
Use only as an emergency treatment, since excessive use may cause secondary effects such as nausea and vomiting. The skin wash and douche produce no secondary effects, however.

Notes:
Gumbo limbo is common in the neotropics—a tree is seemingly always at hand. When you harvest the bark, make sure to cut from branches only (not the trunk) so that you don't kill the tree.

INDIO DESNUDO or JIÑOCUAVE
Bursera simaruba
BURSERACEAE

Hibiscus

AMAPOLA
Hibiscus spp.
MALVACEAE

Geo-distribution:
An ornamental plant from the Old World tropics, hibiscus is now common in the neotropics, where it grows in backyards and patios at lower elevations.

Botanical Description:
This bushy, woody shrub has glossy green, dentate, alternate leaves and bright red flowers.

Medicinal Uses:
In the neotropics, the hibiscus is very popular as a decorative plant. Some people eat the nutritious leaves and flowers or use them as medicine. The leaves make an interesting addition to any salad; the colorful flowers, which are also edible, are used to decorate servings of food. The leaves also provide relief for painful menstruation. Using the leaves and flowers together you can make a

tea that helps staunch excessive menstrual flow and prevent postpartum hemorrhages and miscarriage. To treat skin problems, use the tea as a bath, compress, or maceration. In Belize, some people believe that only the red flowering hibiscus, *H. rosa-sinensis*, has medicinal properties; Costa Ricans, however, categorize both that species and *H. sabdariffa* as medicinal plants.

Nutritive, emollient, antihemorrhagic, emmenagogue, vulnerary.

Preparation:
To prepare tea, boil a handful of leaves and flowers in 1 L of water. Drink 1-3 cups per day.

Notes:
Settlers from the Caribbean Basin introduced *Hibiscus sabdariffa*—called Jamaican hibiscus or roselle—to Costa Rica (see photo, left). This handsome annual plant is related to common hibiscus, okra, and cotton; it is both decorative and useful. You can make a delicious, chilled tea with the dried red calyces of the flowers. Its refreshing citrus flavor and striking red color make it a great natural alternative to the artificial drinks kids seem to crave so much. The leaves of this species are also edible. Hibiscus is generally propagated from woody stem cuttings that are planted directly in permanent sites. Jamaican hibiscus, however, is started from seed.

Hoja de estrella

Geo-distribution:
Native to the neotropics, *hoja de estrella* occurs from Mexico to Central America and parts of South America. This plant, which is very common in Costa Rica, usually grows in moist, shady forest habitats along streams and rivers, from sea level to midelevations.

Botanical Description:
Hoja de estrella is an herbaceous bush that grows 1–5 m tall with a round, thick stem and large, alternate, round leaves 8–31 cm in diameter. The leaves are deeply heart shaped at the base of the stem and finely pubescent on the upper side; they give off a strong aromatic scent similar to the sassafras tree of North America. The small green flowers are borne on delicate spikes opposite the leaves.

Medicinal Uses:
Hoja de estrella should only be used externally, either as a topical maceration to help relieve skin problems, sore muscles, and swelling caused by injuries, or as a bath for aches and pains, sore muscles, rheumatism, and swelling. Folk remedies recommend applying macerated leaves to the forehead to help relieve headaches.

Vulnerary.

Preparation:
To prepare a maceration, crush fresh leaves with your fingers or in a blender. To prepare a sponge bath, boil 2 handfuls of leaves in 4 L of water for about 10 minutes.

HOJA DE ESTRELLA
Piper auritum
PIPERACEA

Caution:
Hoja de estrella should not be taken internally; the aromatic oil in the leaves is considered slightly toxic and may pose a health risk.

Notes:
Many people harvest wild *hoja de estrella*. For best results in your home garden, plant in shady, moist areas; one option is to transplant root cuttings from the wild to your garden.

Horsetail

Geo-distribution:
A primitive plant found in many parts of the world, horsetail occurs at lower elevations in the neotropics. It grows in wet, marshy areas with loamy soils.

Botanical Description:
Horsetail has a stringlike rootstock with roots at the nodes; it produces numerous hollow stems of two types. First, a fertile, beige stem grows out of the ground, reaching a height of 10–20 cm and bearing a canelike spike at the apex that contains spores; this stem quickly dies. Next grows a green, sterile stem (up to 50 cm) that features whorls of small branches.

Medicinal Uses:
This herb has been used worldwide as a diuretic to treat kidney and bladder problems. Horsetail is said to help prevent hair loss, to stimulate nail growth, and to strengthen teeth, bones, and skin—the leaves have high concentrations of calcium and silica. It has also been used as a wash for healing wounds. Traditional uses include treatment of cystitis, intestinal disorders, rheumatism, and gout.

Nutritive, diuretic, astringent.

Preparation:
To make an infusion, steep a handful of fresh leaves in 1 L of water. To make a decoction, boil the same mixture for 10 minutes. Recommended dosage for adults is 1 cup per day for approximately 7 days.

Caution:
Pregnant women and people with heart or kidney disease should not use horsetail.

COLA DE CABALLO
Equisetum spp.
EQUISETACEAE

Indian almond

ALMENDRO
Terminalia catappa
COMBRETACEAE

Geo-distribution:
As its name suggests, the Indian almond tree is native to India. In the neotropics, it has adapted well to lower elevations, particularly along beaches, but it is also found inland up to 600 m above sea level. These trees are popular in yards, orchards, and public squares because of their dense canopies, which provide shade.

Botanical Description:
Indian almond trees can reach up to 24 m in height and 1 m in diameter; average height, however, is closer to 10-12 m. The horizontal branches grow in whorls that spread out from the trunk several meters apart. The leaves (10-30 cm long) are dark green on the superior side and pale green beneath. The white flowers are borne on 5-15 cm-long spikes. The almondlike fruit (4-5 cm long) is an excellent source of nutrients.

Medicinal Uses:
The leaves and bark have been used as decoctions to treat skin rashes and infections, hemorrhoids, and breast sores. An infusion of the leaves and bark is said to be effective in treating diarrhea and dysentery.

Antiseptic, antibacterial, vulnerary, antidiarrheal.

Preparation:
Collect a handful of the fresh leaves and bark from a small branch and boil them in 1 L of water for 20 minutes; let the liquid cool before applying as a skin wash. To prepare an infusion, steep a handful of leaves in 1 L of water. Drink 3 cups per week.

Notes:
In Costa Rica, these trees are an excellent option for home landscaping. In addition to providing delicious almonds and a source of medicine, they also create a shady spot for a hammock. You can either start the seeds in plastic nursery bags or plant them directly to permanent sites. Be sure to provide protective fencing for young trees to prevent them from being trampled.

Jackass bitters

Geo-distribution:
Native to the neotropics, jackass bitters occurs at lower elevations, in clearings, pastures, and at roadsides.

Botanical Description:
Jackass bitters is a tall herb that grows to 1–4 m. The leaves have three distinct points, which give them the appearance of a hawklike bird—hence the Spanish name *gavilana* (*gavilán* means hawk). The plant bears yellow daisylike flowers.

Medicinal Uses:
This bitter-tasting herb is used in emergency situations to help treat fevers, diarrhea, intestinal parasites, and infections. The fresh leaves are used to treat many skin problems, including screwworms and bacterial and fungal infections. The leaves are also used to make a skin wash that treats skin infections, as a hair rinse for lice, and as a douche for leucorrhea.

Febrifuge, stomachic, antidiarrheal, antiparasitic, antibacterial, antifungal, antiviral.

Preparation:
A popular folk remedy calls for applying mashed leaves to the skin. To prepare tea, mix 1 fresh leaf per cup of water and boil for 10 minutes. Take 1–3 cups per day. For external use—as a skin wash, for example—boil a handful of leaves in 4 L of water for 10 minutes.

Notes:
Some gardeners use the tea both as an insect repellant and as a fungicide for plants. Wild seeds can be planted in the home herb garden, preferably in a patch that receives full sun.

GAVILANA
Neuroloena lobata
ASTERACEAE

Juanilama

Geo-distribution:
Juanilama is native to Mesoamerica; in Costa Rica, where it is a popular medicinal herb, it grows in patios and backyards at most elevations.

Botanical Description:
This mintlike aromatic herb's woody stems grow upright to 1 m and carry light-green leaves that are serrate and opposite. White and lavender flowers grow in the leaf apex.

Medicinal Uses:
Use juanilama to make a flavorful mint-tea substitute; it is well-known for its relaxant and digestive properties. In addition to this, the herb has been reported to aid in the treatment of liver problems, menstrual problems, colds, and flu. An alcohol tincture of juanilama treats pain related to arthritis and rheumatism.

Digestive, nervine, antispasmodic, expectorant, sudorific, emmenagogue.

Preparation:
To make an infusion, steep a handful of fresh leaves in 1 L of water (or 1 tablespoon of the dried herb in 1 cup of water) for 10 minutes. To make a tincture for external use, soak 2 handfuls of fresh leaves in 1 L of alcohol (white rum is one option).

Notes:
Juanilama makes an excellent tea. Mix it with lemongrass and ginger for a tasty everyday drink. Propagating new plants from woody stem cuttings is easy: Either plant the stem cuttings in plastic nursery bags with prepared potting soil or plant them directly in permanent sites.

JUANILAMA
Lippia alba
VERBENACEAE

Lemongrass

ZACATE DE LIMÓN
Cymbopogon citratus
POACEAE

Geo-distribution:
Originally from India, lemongrass is now cultivated throughout the neotropics—at all elevations. In Costa Rica, lemongrass is a very popular medicinal plant (and cooking ingredient), and many people grow it in their patios.

Botanical Description:
This grasslike plant grows to 1 m tall and has light-green leaves that give off a lemon scent.

Medicinal Uses:
Lemongrass is a popular folk remedy for colds and flu, gastrointestinal disorders, nervous conditions, pain, and inflammation. *Digestive, relaxant, expectorant, febrifuge.*

Preparation:
To make an infusion, steep a handful of leaves in 1 L of water. Costa Rican campesinos often boil the root and the lower part of the stem to make a stronger decoction. Drink 1–3 cups per day.

Notes:
This medicinal plant is easy to grow in tropical home gardens. To propagate it, simply remove young plants from the base of a mature plant and transplant them.

Life everlasting

Geo-distribution:
In Costa Rica, this common tropical herb occurs in backyards and patios at lower elevations.

Botanical Description:
A succulent herb, life everlasting grows up to 1 m tall and carries shiny green leaves that have scalloped edges. The lantern-shaped flowers grow on tall panicles with shades of purple and pink.

Medicinal Uses:
Although this is generally considered an ornamental plant, it has several interesting medicinal uses. As with aloe vera, the leaves of life everlasting are mashed to extract a juice that serves as treatment for many skin problems, headaches, bruises, strains, and sprains. A decoction or water extract is taken internally to treat colds and flu, chronic illnesses, and infections.

Anodyne, antiherpetic, antiseptic, astringent, hemostatic.

Preparation:
For external treatment, mash the leaves and apply to the skin as a poultice or compress. To make an aqueous solution for internal treatment, add a handful of chopped leaves to 1 L of water. Let the solution stand overnight and drink the liquid throughout the day, as you would water.

Caution:
Studies have found the flowers to be toxic; never use them for medicinal purposes (Arvigo and Balick 1993).

Notes:
This plant is easy to grow and serves as an attractive ornamental in home gardens.

HOJA DEL AIRE
Kalanchoe pinnata
CRASSULACEAE

Mimosa

DORMILONA
Mimosa pudica
MIMOSOIDEAE

Geo-distribution:
Mimosa, which appears in many parts of the neotropics, grows at lower elevations, in pastures, roadsides, and lawns.

Botanical Description:
This low creeping plant has thorny stems and 15-25 pairs of pinnate leaves (5-10 mm long) that fold up when touched. The plant bears fluffy flowers that are a pastel purple.

Medicinal Uses:
Mimosa has a reputation as a sedative, pain reliever, and sleep aid. It has also been used to treat toothaches.

Sedative, anodyne, antispasmodic.

Preparation:
Boil a handful of the leaves in 1 L of water. Sip 3-6 cups throughout the day instead of water. Boil the root in water to produce a concentrate from which you can make a poultice for toothaches. To alleviate nervous problems and insomnia, add powder made from the leaves to food (to prepare the powder, dry the leaves in an oven and then grind in a mortar).

Mint

MENTA DE PALO
Satureja viminea
LAMIACEAE

Geo-distribution:
Mint, or *menta de palo*, is native to the neotropics. In Costa Rica, this hardy and bushy plant occurs at most elevations (you'll often see it in home patios). This species should not be confused with peppermint and spearmint, which are also called "mint."

Botanical description:
Satureja viminea is a shrublike perennial with woody stems. It carries ovate, opposite leaves without serrate edges. The minute flowers are purplish lavender. The plant gives off a strong mint aroma.

Medicinal Uses:
The mint plants, including *S. viminea*, compose perhaps one of the most important groups of medicinal herbs. Menthol, the essential oil, aids in digestion, calms nerves, fights bacteria, and prevents unwanted microbes from colonizing the digestive tract. Regular use of mint oil as a mouthwash helps prevent cavities. Most mints have low toxicity, so its OK to drink hot and cold mint drinks on a fairly regular basis. Note, however, that the essential oil in *S. viminea* is stronger than that found in most other mints and can cause nausea if overused. This herb is appropriate in emergency situations—it is sometimes used to diminish the involuntary contractions of patients who are vomiting, and also to treat indigestion or insomnia. The author recommends peppermint and spearmint for everyday consumption (see p. 73).

Carminative, stomachic, antibacterial, relaxant.

Preparation:
To preserve the essential oils, prepare an infusion or steep the leaves in water, preferably in a closed teapot. A handful of fresh leaves or a tablespoon or two of dried mint leaves per liter of water makes a good tasting tea. You can add honey or dried stevia leaves to sweeten the tea.

Notes:
Mints are easy to grow in the home garden. *S. viminea* is one of Costa Rica's most popular home garden herbs. It can be propagated from woody stem cuttings that are planted directly in the ground or in pots with prepared potting soil.

Mozote de caballo

Geo-distribution:
In Costa Rica, this neotropical native usually grows in pastures, along riverbanks, and near roads, from low- to midelevations.

Botanical Description:
An annual herbaceous plant, the *mozote de caballo* has woody stems that grow to 1-1.5 m in height. It bears alternate, hairy, serrate leaves with three lobes. The small, yellow flowers are borne on long panicles, which form spiny seed capsules. These seed capsules stick to clothes and animals—hence the Spanish name, *mozote de caballo*, which translates loosely as "horse sticker."

MOZOTE DE CABALLO
Triumfetta semitriloba
TILIACEAE

Medicinal Uses:
The cambium layer of the woody stems contains a mucilaginous substance used to treat gastritis, ulcers, diarrheas, dysentery, kidney problems, colds, and flu. Much like aloe vera juice, this remedy soothes the stomach and intestinal lining, and acts as an expectorant and diuretic.

Emollient, digestive, laxative, expectorant, diuretic.

Preparation:
Scrape the bark from a 15-30 cm-long segment of the woody stem. Soak in 1 L of water overnight or steep in hot water to prepare an infusion. Use the mucilaginous solution to prepare a warm tea or to make a cold drink, known as *fresco de mozote* in Spanish. Drink 3 cups per day. *Mozote de caballo* is non-toxic and can be used safely for extended periods.

Notes:
This herb thrives in the home garden. Collect seeds from mature plants and plant them 2-5 cm deep in plastic nursery bags with moderately fertile soil. Keep the soil moist until the seeds germinate, then water twice a week. When the seedlings are about 30 cm tall, transplant them to permanent sites, preferably at a distance from the main garden and house, since the burrs from mature plants are bothersome when they cling to your clothes. Costa Rican campesinos use the stem's mucilaginous substance when making *tapa dulce* (brown sugar). A bucket of *mozote de caballo* is added to the boiling sugar cane juice to clean it of impurities. This process is much healthier than a chemical process that relies on aluminum bisulfate, a known carcinogenic.

Noni

NONI
Morinda citrifolia
RUBIACEAE

Geo-distribution:
Noni (or Indian mulberry) is native to the islands of Polynesia, but has been growing in Central America for quite some time; Asian immigrants brought the plant to Panama when they arrived to work on construction of the canal. Since then, noni has been propagated as a medicinal plant in home gardens along the coastal regions of Central America.

Description:
Noni is a small shrub or tree, usually less than 3 m tall, occasionally up to 6 m. It has glossy, evergreen leaves and yellowish white flowers. The oblong fruit is 10–15 cm long with distinctive scales on the surface. It has an unappealing odor when ripe.

Medicinal Uses:
Noni is a traditional Hawaiian medicinal plant, reportedly used as a tonic for a wide range of illnesses. The ripe fruit is sometimes used to make a poultice for facial blemishes—the poultice is rubbed over the surface of the skin to absorb facial oil (or, in the case of a staff infection, for example, to draw out the pus from an infected sore or boil). The fruit is used to relieve symptoms of arthritis, diabetes,

rheumatism, and the chronic illnesses of old age; the fruit juice has been used to remove head lice. To staunch bleeding from deep cuts, carefully place the young unripe fruit over the cut. Pound the leaves and bark of the stem and then strain them to make a tonic that treats muscle and joint pain and urinary disorders (Etkin and McMillen 2003).

Tonic, antiherpetic, antiseptic, stomachic, depurative, pectoral, antiarthritic.

Preparation:

There are two standard recipes, both of which produce similar results. The simplest consists of blending and straining the soft, ripened fruit and then mixing the resulting liquid with water and either pineapple or citrus juice. Drink several shot glasses of the juice per day (the juice will keep in the refrigerator for up to two weeks). The second recipe calls for filling a widemouthed glass jar with the fruit. Cap the jar tightly and set it in a sunny spot for several days until the fruit turns to mush and the sun-cooked juices drain into the bottom of the jar. Strain the juice into a smaller jar, then refrigerate before using.

Notes:

Because of the recent popularity of noni juice as a natural health drink, the fruits are now available in many Costa Rican markets. Collect the seeds and plant them 2–3 cm deep in plastic nursery bags or containers with moderately fertile soil. Place the containers in the shade; water frequently for one month or until the seeds sprout; when the seedlings sprout, move the containers to an area that receives sunlight and water them twice a week; three to six months later, the seedlings are ready to be planted in permanent sites. The tree does well in a wide range of soils and produces fruit in three to five years. No significant insect problems are reported with noni trees.

Oregano

ORÉGANO
Lippia graveolens and
Origanum vulgare
VERBENACEAE and LABIATAE

Geo-distribution:
You will find two distinct species in the tropics that are called oregano. The most common—*Lippia graveolens* (shown in photo above)—is native to the neotropics and grows in many backyards and patios. The other species, *Origanum vulgare*, is native to Europe but was brought by early settlers to Costa Rica, where it grows at higher elevations.

Botanical Description:
Native oregano is a woody-stemmed, shrublike perennial. The plant has square stems and small, ovate, lanceolate leaves that have serrate edges. The leaves give off a strong oregano scent. The plant's tiny flowers are yellowish white.

Medicinal Uses:
Both types of oregano are generally used as condiments, but they also have some medicinal properties. Prepare an infusion to relieve nervous headaches, upset stomach, indigestion, colds and flu, colic, and coughs. Some women also take an infusion during the three days before their menstrual period to insure regularity and ease premenstrual symptoms. According to the companies that produce it, the commercial essential oil is supposed to aid in the treatment of skin problems.

Stimulant, carminative, diaphoretic, tonic.

Preparation:
Fresh oregano is a wonderful addition to any salad dressing, and the dried herb is excellent in cooked dishes. Steep one handful of fresh oregano or 1 tablespoon of dried herb in 1 L of water for 10 minutes. Take up to 3 cups per day. The essential oil can be purchased in most health food stores.

Notes:
Oregano is a hardy herb and ideal for growing in the tropical home garden. With a few plants around, you will rarely need to buy it at the store. Propagate new plants from the woody-stem cuttings of mature plants. Dried oregano should be placed in an airtight container to prevent molding.

Papaya

Geo-distribution:
Native to the neotropics, papaya grows in tropical areas around the world. In Costa Rica, it thrives at lower elevations.

Botanical Description:
Papaya is actually an herbaceous perennial plant and not a tree. The pulpy stem has deep, palmate leaves on long stems. The papaya's flowers are greenish yellow. The large, ovate fruits grow up to 40 cm long; they are yellowish orange when ripe.

Medicinal Uses:
The papaya fruit is an exemplary case of food that can be used as medicine. It is effective in treating a wide range of digestive problems—indigestion and constipation, for example; it is also used to treat liver problems and to lower high blood pressure; and, it is used as a diuretic for the kidneys. The medicinal properties of the fruit are enhanced when it is mixed with aloe vera juice. You can use the rind of the fruit to heal skin problems or simply as a beauty treatment. The sap of the trunk is said to be useful in treating warts.

Nutritive, stomachic, laxative, hepatic, diuretic, depurative, anti-malarial.

Preparation:
To make healthful, tasty papaya-and-aloe juice, fillet one large aloe leaf. In a blender, combine the aloe pulp and a peeled, seedless papaya in 1 L of water; add more papaya to taste. Drink 3 cups per day between meals.

Notes:
In Costa Rica, papaya thrives at lower elevations. On a mere two square meters of land, one plant can produce up to 68 kg of fruit per year!

PAPAYA
Carica papaya
CARICACEAE

Parsley

Geo-distribution:
Originally from Europe, parsley was introduced to the neotropics by the early settlers. In Costa Rica, it grows well at higher elevations.

Botanical Description:
This biennial herb has glabrous, opposite, dentate leaves and umbels of small yellow flowers.

Medicinal Uses:
Most people think of parsley as a spice, garnish, or salad green, but it also has some surprising medicinal uses. Parsley root is used as a potent diuretic for the kidney and bladder, and a tea made from the seeds helps regulate menstruation. People have also discovered the healing power of parsley juice—a wonderful tonic for the body. Traditionally, parsley has been used to treat jaundice, asthma, coughs, suppressed or difficult menstruation, obesity, bed-wetting, rheumatism, tumors, edema, indigestion, worms, and kidney stones (Lust 1974).

Antispasmodic, carminative, diuretic, emmenagogue, expectorant.

Preparation:
Adding lots of parsley to salads is a good way to help prevent health problems. Parsley juice, which contains several minerals and vitamins, is particularly beneficial for rejuvenation and healing. If you grow your own parsley, you can collect the roots at the end of the growing season and follow this recipe:

1. Wash the roots thoroughly.
2. Chop into small pieces.

PEREJIL
Petroselinium crispum
UMBELLIFERACEAE

3. Dry the pieces thoroughly and store them in an airtight container.
4. Use 1–2 g of dried, powdered parsley root per cup of hot water.
5. Drink 3 cups per day until results are obtained.

Notes:
Parsley is easy to cultivate in home gardens at higher elevations. Plants are started from seed. A dozen plants can keep a family well supplied with parsley greens and juice for optimal health.

Passion flower

GRANADILLA REAL
Passiflora quadrangularis
PASSIFLORACEAE

Geo-distribution:
Native to Brazil, this passion flower species produces a delicious fruit. In Costa Rica, it is cultivated at higher elevations.

Botanical Description:
The plant is a woody, hairy climbing vine with alternate, serrate leaves. The flowers are beige with purple tints. The ovoid fruit is a berry about 5–8 cm long with many seeds and a wonderful taste.

Medicinal Uses:
Passion flower is used as an emergency herb for nervous disorders, particularly insomnia, hysteria, depressions, and nervous headaches. The ground seeds are reportedly used as a deworming agent. *Sedative.*

Preparation:
A tincture is one of the best ways to administer this medicine (see the Caution). Half-fill a glass container with leaves and then top off with vodka. Seal the container and shake the bottle vigorously once a day for three weeks. Decant the liquid and store in a dark bottle for future use. Use 15–60 drops per day for approximately one week.

Caution:
Passion flower tincture is slightly toxic and a strong sedative; use sparingly to avoid overdoses.

Peppermint and Spearmint

Geo-distribution:
Most mint species were brought from the Old World to Costa Rica by settlers. In Costa Rica, they grow at high elevations, in moist habitats. Only *Satureja viminea* is native to the tropics (see Mint, p. 64).

Botanical Description:
Mentha species are perennial low-growing herbs that reach 30 cm tall; they have running tendrils, a strong mint aroma, and a square stem. The leaves are dark green, lance-shaped, sharply toothed, and smooth (often with hairs). The flowers are small and purplish and have a tubular 5-toothed calyx and a 4-lobed corolla.

Medicinal Uses:
The Spanish common name for these species is *hierbabuena*, which means "good herb"—and that is exactly what it is. In fact, of all the medicinal herbs, the mints are perhaps of the greatest medicinal value. They contain menthol, an essential oil that acts as a nerve calming agent and that also improves digestion; in addition, it is antibacterial and prevents unwanted microbes from living in the digestive tract. Regular use of mint oil as a mouthwash helps prevent cavities. Most mints are gentle and non-toxic, so they can be served frequently as hot and cold drinks for the family. Traditionally, various mints have

HIERBABUENA
Mentha spp.
LABIATAE

been used to relieve insomnia, cramps, gas, coughs, headache, heartburn, nausea, abdominal pains, and other problems related to nerves.

Carminative, stomachic, expectorant, relaxant, antibacterial.

Preparation:
All mints are best prepared by infusion or steeping, preferably in a closed teapot to preserve the essential oils. To make a tea, steep a handful of fresh leaves or a tablespoon or two of dried leaves in 1 L of water. Honey or dried stevia leaves are perfect sweeteners.

Notes:
Mints are easy to grow in home gardens. To propagate, select vegetative offshoots or runners and plant directly in permanent sites.

Pink shower tree

Geo-distribution:
The pink shower tree is native to the neotropics. In Costa Rica, it grows primarily in forests and open fields at lower elevations; it is often planted around homes either as an ornamental or for its medicinal properties.

Botanical Description:
The pink shower tree, which has a distinctive spreading crown, grows to a height of 30 m. The leaves are pinnate with 10–20 pairs of 3–5 cm-long leaflets. During the dry season, the tree sheds its old leaves, giving way to beautiful racemes of pastel pink flowers. The dark-brown wood-like seed capsules reach lengths of up to 50 cm and have many seeds. These seeds are separated by resinous membranes that taste somewhat like carob.

Medicinal Uses:
The dark, caroblike syrup from the seedpod makes a nutritious drink (rich in vitamins and minerals) that is used to treat anemia and a general state of lethargy. The leaves act as a laxative and diuretic to treat constipation, digestive problems, edema, backaches, and kidney problems. Fresh leaf juice is used to treat skin problems, particularly fungal infections such as ringworm.

Nutritive, laxative, diuretic, antifungal.

Preparation:
To extract the syrup, boil several seed-capsule membranes in water or milk. Strain the liquid and drink 1 cup per day. To treat skin problems, macerate fresh leaves to a juicy pulp and apply the pulp to the skin several times a day until results are obtained. To prepare a tonic, boil a handful of the leaves in 1 L of water; drink throughout the day. To prepare a laxative, boil half a cup of leaves in a cup of water for 10–20 minutes; strain before serving. Drink 1 cup per day.

Notes:
This is a beautiful ornamental. Start the seeds in cups, pots, or plastic nursery bags with moderately fertile soil. The seedlings grow quickly and are quite hardy—you don't need to do much to keep them healthy. Because it grows to such a large size, this tree is not recommended for small home sites.

CARAO
Cassia grandis
CAESALPINIOIDEAE

Pink trumpet tree and Pau d'arco

ROBLE DE SABANA
and PAU D'ARCO
Tabebuia rosea and
T. impetiginosa
BIGNONIACEAE

Geo-distribution:
Both of these stunning trees are native to the neotropics. Pau d'arco (*T. impetiginosa*)—more common in South America than it is in Central America—has several common names, including *lapacho*, *taheebo*, *ipe*, and *cortez negro*, as it is called in Costa Rica. In Mesoamerica you are more likely to see the pink trumpet tree (*T. rosea*, in photo above) than you are to see pau d'arco. Also known as *roble colorado*, this tree is the national tree of El Salvador. In part because these trees are a favorite choice of landscape designers, both are now pantropical; indeed, throughout the tropics, you will see them in parks and along city avenues. In Costa Rica, they are commonly found at lower elevations, in both pastures and savannas.

Botanical Description:
The flowers of the *T. impetiginosa* are pastel scarlet and purple. The trunk is dark gray and the inner bark has black stains. The leaves on the younger trees have serrate edges. The flowers of *T. rosea* (formerly *T. pentaphylla*) range from pastel pink to pastel maroon. Its trunk has gray bark; the leaves are compound, without serrated

edges. Both of these species lose their leaves in the dry season.

Medicinal Uses:

For centuries, people in the neotropics have used the inner bark of both species to make folk remedies. (The Maya, the Inca, and other peoples venerated these trees as sacred.) The bark, a powerful botanical medicine that restores the immune system, has been cited traditionally to treat the following health problems: anemia, arteriosclerosis, arthritis, asthma, bronchitis, boils, cancer, candida, colitis, colds, constipation, coughs, cystitis, diabetes, diarrhea, emphysema, dysentery, eczema, fevers, flu, gastritis, gallbladder problems, infections, liver problems, lung problems, leukemia, pain relief, parasites, prostatitis, pyorrhea, and wounds. No wonder the native indigenous people considered it a sacred tree! After fifteen years of successfully treating patients with this bark, the author too considers these majestic trees a sacred gift of nature.

A great deal of research has been done on the medicinal properties of both trees. Researchers have demonstrated that eighteen of the bark's many organic compounds have medicinal properties. These include quinines; compounds that effectively cleanse the human body of common tropical microbes and protozoans; and lapachol, an organic alcohol that is responsible for the bark's tumor-reducing and anti-cancer properties. The trees' tannins are antiseptic and antibacterial. The organic chemistry of these compounds is quite complex; the author recommends reading *Pau d'Arco: Immune Power from the Rainforest*, by Kenneth Jones, p. 66-100.

Preparation:

From mature branches, carefully cut and remove the outer bark down to the cambium layer or cortex. Next, slice off the cortex down to the wood of the branch. Cut these strips of inner bark into 2-3 cm pieces on a clean cutting board and then dry out the pieces quickly (within three days) in a hot, dry place. Store the processed bark in an airtight container. To make a slightly bitter, woody-tasting tea, boil 1-2 g of the bark in 1 cup of water for 10-15 minutes. Drink 1-3 cups per week for preventive health care. For acute or chronic illness use 1-6 cups per day until results are obtained. To make capsules, grind the bark in a blender and place the dry powder in empty gelatin capsules. Take 3-6 capsules per day for acute and chronic illnesses. To make a tincture for external and internal use, add 60 g of powder to 1 L of vodka. Shake once a day and store in a dark place for at least 1 week (some say for at least a month). Use several teaspoons of the tincture per week to prevent health problems, or 3-6 teaspoons per day in the case of chronic or acute illness. This tincture also serves as an effective gargle for a sore throat or as an external treatment for skin problems.

Notes:

Although pau d'arco is sometimes grown in reforestation projects, logging activities threaten the existence of this species in its natural habitats. You can help conservation efforts by starting seedling trees at home and then planting them at permanent sites.

Pisabed

Geo-distribution:
Pisabed occurs in tropical climates around the world. In Costa Rica, it commonly grows in pastures, fields, and along roadsides, both in coastal regions and at midelevations. You will sometimes see it in patios, where it is grown as a medicinal plant. Among the English-speaking Caribbean community of Costa Rica, this plant is frequently called pisabed, though in other places it is also referred to as *frijolillo*, *frijolillo negro*, *sonajera*, and *bajaro*.

Botanical Description:
Pisabed is an annual plant (in areas without frost, it is sometimes a short-lived perennial) with woody stems that often grow to 1–2 m in height. The bushlike plant has 8–12 oval-shaped, compound, pinnate leaflets. The yellow flowers are borne on racemes and produce seed capsules that are light green, smooth, and slightly arched, much like the beak of a bird (the Spanish name of this plant, *pico de pájaro*, literally means bird beak). The seed capsules have two rows of 15–20 dark brown seeds.

Medicinal Uses:
The leaves are used externally to treat skin problems such as inflammations, infections, and fungal rashes; tests have shown that the leaves have antibiotic properties (Robineau 2005). Leaves and capsules were once administered internally for fevers, diarrhea, jaundice, rheumatism, and kidney and bladder dysfunctions (see the Caution). The root has been used to treat colic spasms; in strong doses it acts as a purgative.

Hepatic, antidiarrheal, anti-inflammatory, vulnerary, febrifuge, antispasmodic, antirheumatic, depurative, purgative, anti-influenza.

Preparation:
The fresh leaves can be used as a maceration to treat skin conditions. Apply the freshly mashed or blended leaves directly to the affected area 2-3 times per day. Traditionally, children were often given pisabed tea during the day to prevent them from wetting the bed. To make a tea for intestinal cramps, wash the root thoroughly and chop it into small pieces. Boil a handful of these chips in 1 L of water; drink throughout the day.

Caution:
The seed capsules have tested toxic for internal and external use. Although tea made from fresh leaves is considered safe, pregnant women should avoid drinking it (Robineau 2005).

Notes:
Costa Rican farmers grow this plant in the patio both as an ornamental and as a medicinal herb. To propagate pisabed, collect seeds from a plant and dry them; then plant the seeds directly in garden beds. Alternatively, you can plant the seeds directly in a germination flat, wait until the seedlings reach about 5 cm tall, then transplant them to the garden. The latter method provides the young, tender seedlings with better protection from insects and tropical rains.

PICO DE PÁJARO
Senna occidentalis
CAESALPINIOIDEAE

Pokeweed

Geo-distribution:
Pokeweed occurs in North and Central America, usually in pastures, along streams, and on the edge of forests.

Botanical Description:
The plant can grow into a 1.5-m-tall shrublike bush with shiny, green alternate leaves and reddish purple stems. Spikes up to 50 cm long bear white flowers that produce dark purple fruit.

Medicinal Uses:
People used to use young leaves to make salads; the older leaves were cooked thoroughly (thus extracting the high levels of oxalic acid) to make a spinachlike dish. Although the stems and root are toxic, they can be used to treat head and body lice.

Nutritive, antiparasitic.

Preparation:
Chop up 200 g of the root and boil the pieces in 4 L of water for 30 minutes. Strain the liquid, let it cool, and use it as a sponge bath for the scalp or skin. Repeat for several days until results are obtained.

Caution:
Use pokeweed for external purposes only—internal use may cause serious secondary effects such as nausea and vomiting.

Notes:
Pokeweed is an attractive addition to home gardens, and you can use the berries to make a dark fabric dye. To propogate the plant, collect seeds from older plants and sow them around the home. These plants take up a lot of room, so it is best to keep them at a distance from flowerbeds.

JABONCILLO
Phytolacca decandra and *P. rugosa*
PHYTOLACCACEAE

Prickly pear cactus

TUNA
Opuntia tuna and
Nopalea cochenillifera
CACTACEAE

Geo-distribution:
People use the common name, prickly pear cactus, to refer to several species. Native to Mexico, *Opuntia tuna* and *Nopalea cochenillifera* cacti grow at low elevations in the neotropics. In Costa Rica, they are commonly seen in patios.

Botanical Description:
These plants have rounded stem segments that are covered in spines; yellow or red flowers; and edible red fruits that are covered with delicate spines.

Medicinal Uses:
These cacti have been used for centuries by Mesoamerican indigenous tribes. Much like aloe vera, they have a mucilaginous pulp with many medicinal uses. The pulp is used to treat skin disorders and digestive problems, and it also serves as an insect repellant. In addition, it is said to help in cases of arthritis and rheumatism.

Emollient, stomachic, laxative, antiarthritic, culinary.

Preparation:
Carefully peel the rind from the stem segments and chop the inner pulp into small squares. Soak the pulp in water overnight. In the morning, the liquid will be very mucilaginous. Drink 3–4 glasses per day between meals. Mexicans use the chopped pulp, which they call nopal, to make a popular dish. To prepare nopal, peel the stem segments, cut the pulp into squares, then coat the pulp with bread crumbs and bake.

Notes:
These plants thrive in tropical home gardens. To propagate new plants, simply plant a stem cutting in the ground.

Medicinal Plants of Costa Rica / **81**

Purple mombin

Geo-distribution:
Purple mombin trees are native to the neotropics; in Central America, they grow at low- to midelevations (up to 1,200 m). These trees are rarely found in primary forests; rather, they tend to grow around farms, fields, and homes. This is explained by the fact that the seeds are sterile and the tree is propagated primarily through the agency of humans, who plant branch cuttings. Some researchers believe the tree originated in Mexico and was later distributed throughout the region by indigenous tribes.

Botanical Description:
The purple mombin tree grows to 4–5 m in height. It has a smooth, gray trunk and compound, pinnate leaves with 10 pairs of leaflets. The pastel purple flowers are borne on a raceme and produce edible, oval, oblong drupe fruit (yellow or red). You might well categorize this tasty fruit as medicinal since it is high in vitamin A, vitamin C, iron, phosphorus, and copper.

Medicinal Uses:
Mexicans use the fruit as a diuretic and antispasmodic; they also make a decoction to bathe wounds and heal mouth sores. Syrup prepared from the fruit is taken to overcome chronic diarrhea. An astringent bark decoction is a remedy for mange, ulcers, dysentery, and bloating in infants caused by intestinal gas. The juice of the fresh leaves is a remedy for thrush; a decoction of the leaves and bark is employed as a febrifuge; an infusion of shredded leaves is used for washing cuts, sores, and burns. Researchers have found that an aqueous extract of the leaves has antibacterial properties, and an alcoholic extract is even more effective in this regard. The gum resin of the tree can be blended with pineapple or soursop juice to treat jaundice. By deduction, many of the other uses of the plant suggest that the fruit, leaves, and bark are fairly rich in tannin (Morton 1987).

Nutritive, diuretic, antispasmodic, antidiarrheal, antiherpetic.

Notes:
These useful fruit trees are a nice addition to the home garden. Propagate them by planting 1.5-m-long branch cuttings 30–40 cm deep in permanent sites. One tree usually provides a sufficient source of medicine for an average-size family.

JOCOTE
Spondias purpurea
ANACARDIACEAE

Rat killer tree

MADERO NEGRO
Gliricidia sepium
PAPILIONOIDEAE

Geo-distribution:
The rat killer tree is native to the neotropics; in Costa Rica, it grows at low- to midelevations. It is commonly planted as a living fence and you will see it along roadsides and bordering coffee and cacao plantations.

Medicinal Uses:
The leaves have been used in the treatment of skin problems and body lice. The flowers are edible and have been used to treat diarrhea.

Preparation:
Boil a handful of leaves in 1 L of water for a sponge bath. Use the liquid or a maceration of fresh leaves to help relieve rashes, insect bites, boils, pimples, and burns. An infusion of a handful of flowers in 1 L of water can be used to treat diarrhea. The flowers are sometimes cooked with eggs and onions.

Antiherpetic, nutritive, antidiarrheal.

Notes:
As indicated by its English name, this tree has a particular folk use. Before the advent of modern formulas for killing rodents, the leaves of this tree were boiled along with corn, then dried and used as rodent bait (an active compound in the leaves acts as an enzyme inhibitor in rats and mice). Its scientific name, *Gliricidia*, comes from an Afro-Caribbean English word that means "rat killer."

Red targua

TARGUÁ COLORADO
Croton gossypifolius
EUPHORBIACEAE

Geo-distribution:
Native to the neotropics, red targua occurs in Mexico, Central America, the West Indies, and the northern part of South America. It usually grows at low- to midelevations, often in coffee plantations, second growth, and disturbed areas of the rainforest.

Botanical Description:
This medium-size tree (it reaches 15 m) has smooth, gray bark and alternate, heart shaped leaves (30–40 cm long). Long racemes bear small white flowers, as well as small (5–6 mm) 3-lobed seed capsules. When the bark is cut, the tree exudes an amber-red resinous sap. Several other species of targua exude a red latex sap with similar properties, including *C. costarricensis*, *C. jimenezii*, *C. panamensis*, and *C. xalapensis*.

Medicinal Uses:
The resinous, astringent, and antiseptic sap is used for treating gum diseases such as pyorrhea. In colonial times, it was used as a popular "toothpaste." Apply the sap to minor wounds to stanch bleeding and prevent infections.

Astringent, antiseptic, antibacterial, dentifrice, hemostatic.

Preparation:
Make an incision several centimeters long in the trunk of the tree. As the bark begins to "bleed," collect the sap with a spoon at the base of the incision and store it in a small jar. The sap is best applied to the gums with a cotton swab, preferably in the morning and evening, after brushing your teeth.

Caution:
This medicinal sap is used externally only, since it is slightly toxic (Núñez 1975).

Notes:
Because of the size of this tree, it is not recommended for the home garden. Some Costa Rican markets sell small jars of red targua sap.

Redhead

Geo-distribution:
Redhead—known in Spanish as *zorrillo real*, *palo camarón*, and *coralillo*—is native to Central America and the Caribbean Basin. This plant grows wild at low elevations along roadsides, in pastures, and secondary growth; it is planted in backyards as an ornamental.

Botanical Description:
This bushlike tree grows up to 2 m tall with ternate, lance-oblong to rounded-ovate leaves that are 5–21 cm long. The redhead's inflorescences bear many reddish orange flowers. The small fruits (only 6–10 mm long) are said to be edible.

Medicinal Uses:
Traditionally, this plant was used to remedy a wide array of skin problems, including sores, bruises, rashes, itches, insect bites or stings, burns, cuts, and fungal infections.

Recent research has demonstrated that the active compounds in the leaves have antibacterial and antifungal properties and also display analgesic activity (Robineau 2005).

Vulnerary.

Preparation:
In order to treat skin problems, make a skin wash or bath: Boil several handfuls of leaves and flowers in 4 L of water until 2 L of con-

ZORRILLO REAL
Hamelia patens
RUBIACEAE

centrate remain, and then strain. In Belize, juice from the crushed leaves is applied to insect stings and bot fly parasites that have burrowed into human flesh. In Haiti, the fresh, mashed leaves are applied to the forehead to alleviate migraines.

Caution:
Use externally only. According to the UNESCO-funded TRAMIL 3 Workshop on Central American and Caribbean medicinal plants, laboratory studies indicated that this plant has toxic effects if taken internally (Robineau 2005).

Notes:
Redhead makes an attractive ornamental for the home garden. To propagate new plants, prune stem cuttings from mature shrubs; plant the cuttings in pots filled with prepared potting soil and place them in a shady area; as soon as the cuttings begin to sprout new leaves, transplant them to permanent sites (if you wait too long, the cuttings will become root-bound). Prune regularly if you want these plants to grow as compact shrubs in your garden. Redhead plants are hardy—you don't need to water them during the dry season.

Rose periwinkle

Geo-distribution:
Originally from Madagascar, rose periwinkle is now naturalized in tropical climates around the world. In Costa Rica, it is a common ornamental plant in patios and public places.

Botanical Description:
The flowers, which bloom most of the year, come in many colors—pure white, pastel rose, violet, red, and orange. The glossy, evergreen, opposite leaves have distinctive white veins.

Medicinal Uses:
Rose periwinkle has been used for centuries throughout the Americas to treat high blood pressure, asthma, constipation, and menstrual problems (and to relieve symptoms of diabetes). In Costa Rica, it has been used as a blood purifier and to treat sore throats, toothaches, intestinal parasites, and hemorrhages.

Studies over the past 20 years have uncovered a great deal of information on this plant's biochemistry. Rose periwinkle contains many alkaloids, some useful, some toxic. Two of its most valuable compounds—vincristine and vinblastine—are now extracted to make medications for treating Hodgkin's disease, various kinds of leukemia, and skin, lymph, and breast cancer. Companies in Australia, Africa, India, and southern Europe now grow periwinkles commercially for medicinal use. Whatever the medicinal value of some plant compounds, however, research has also demonstrated that most of the 72 components from the plant are toxic and can cause kidney and nervous system disorders if taken internally (Westbrooks and Preacher 1986).

MARIPOSILLA
Catharanthus roseus
APOCYNACEAE

Preparation:
See the Caution.

Caution:
Continued ingestion of this plant can have harmful secondary effects due to its toxicity. The information above is included for its historical ethnobotanical value only; the plant should not be used for home medicinal purposes.

Rosemary

Geo-distribution:
European colonists introduced this plant to the neotropics. In Costa Rica, it thrives at higher elevations and is also a favorite in home gardens.

Botanical Description:
Rosemary is a perennial woody-stemmed herb with ash-colored scaly bark and many branches. The herb has opposite, leathery, thick leaves, which are a lustrous dark green above and a downy white underneath. The pale-blue flowers (sometimes white) grow in short, axial racemes.

ROMERO
Rosmarinus officinalis
LABIATAE

Medicinal Uses:
The aromatic oil in rosemary promotes digestion, stimulates bile, and improves liver function and circulation. Lightly season salad dressings and vegetable dishes with rosemary to help keep the digestive system functioning smoothly. The essential oil is used as a massage oil to soothe sore muscles, improve lymphatic circulation, and stimulate sweating. An infusion is often used as a rinse to improve the health of both scalp and hair.

Antispasmodic, astringent, emmenagogue, stimulant, stomachic, sudorific, tonic.

Preparation:
To make an infusion, steep 1 tablespoon of fresh rosemary in 1 L of water for 10 minutes. Sip a maximum of 1 cup per day. To make a hair rinse, steep a handful of rosemary in 1 L of water. To prepare a tincture to relieve muscular pain and skin infections: Add a handful of rosemary leaves to 250 mL of clear rum or vodka in a glass bottle with a tight seal; shake daily for 7 days before using.

Caution:
Excessive use of rosemary may cause secondary effects such as nausea and vomiting. Use with caution as a medicinal herb.

Notes:
You can find rosemary at many Costa Rican herb stands. To propagate new plants in your home herb garden, plant fresh cuttings in pots filled with sandy soil; water the cuttings regularly until they root; and transplant them to permanent sites.

Rue

RUDA
Ruta graveolens
RUTACEAE

Geo-distribution:
European immigrants transported rue to the New World. Today, in Costa Rica, it is commonly grown in patios with other favorite herbs such as mint and oregano.

Botanical Description:
Rue is an herbaceous perennial plant with unusual delicate, pinnate, blue green leaves (and stems of the same color). The yellow flowers have five petals. The leaves give off one of the most peculiar scents in the plant kingdom; the scent is very aromatic and somewhat narcotic.

Medicinal Uses:
Note: Although this plant was used traditionally for medicinal purposes, investigations have cast serious doubts on its medicinal value (Lust 1974).

Rue was used for centuries as a popular folk remedy in both Europe and the Americas; traditional applications included treatment for painful (or late) menstruation, stomach problems, vomiting, headaches, earaches, and intestinal parasites. It was also used to ease childbirth pains and to calm the nerves, and it has even been used as an abortive.

Preparation:
This plant should not be used medicinally (see the Caution).

Caution:
Numerous studies have indicated that the powerful aromatic oil in rue is toxic and may cause serious secondary effects, including nausea, vomiting, stomach cramps, dizziness, delirium, tremors, and even unconsciousness. Used as an abortive, it may be fatal for the mother. Applied to the skin it may cause rashes and blistering (Lust 1974).

Sage

Geo-distribution:
Originally from Europe and first introduced to the neotropics by early settlers, sage now grows at higher elevations in Costa Rica, where it is often planted in patio herb gardens.

Botanical Description:
Sage is a 30–60 cm-tall herb with wiry stems and purplish flowers. The grayish-green leaves are opposite and finely wrinkled with oblong, rounded ends.

Medicinal Uses:
Although this herb is used primarily as a seasoning, sage tea is also used to treat nervous conditions, digestive disorders, an assortment of female maladies, and fevers.

Astringent, antispasmodic, diaphoretic, carminative, nervine, tonic.

Preparation:
To make an infusion, steep a handful of fresh leaves or a tablespoon of dried sage in 1 L of water. Drink 1-2 cups per day.

Caution:
The aromatic essential oil is strong; overuse may cause nausea and vomiting.

Notes:
This useful herb thrives in the home garden. Many Costa Rican herb stands and markets sell live plants or fresh cuttings, which are used to propagate new plants. Plant the woody stem sections in pots with prepared potting soil. Place the pots in the shade and keep the soil moist until the plants are well established; then transplant into the garden.

SALVIA
Salvia officinales
LABIATAE

Sansevieria

Geo-distribution:
Sansevieria, a common houseplant found worldwide, is used by people in the neotropics to make fence rows.

Botanical Description:
This is an upright perennial plant with stiff, smooth vertical leaves that are lanceolate (50 cm tall); the leaves are variegated in color, combining both light green and dark green. The greenish white flowers, which occur in racemes, give off a fragrant scent.

Medicinal Uses:
This plant is associated with a number of tropical legends and folk remedies. In many regions, it is considered a useful treatment for venomous snakebites and skin conditions such as sores and rashes. Indigenous communities in South America use it principally for skin problems, and tests have demonstrated that the juice has antiviral properties (Schultes and Raffauf 1995).

Preparation:
Sansevieria is most commonly applied as a maceration for skin problems. In Belize, people chew on the leaves as a remedy for venomous snake bites (Arvigo and Balick 1993). To prepare an aqueous solution, chop a handful of leaves into small pieces and soak them in 1 L of water overnight. The solution is either drunk or used as a skin wash—much like aloe vera. Some campesinos feed this solution to their chickens to keep them healthy.

Antifungal, astringent, antiseptic, antibacterial, antiherpetic.

LENGUA DE SUEGRA
Sansevieria trifasciata
AGAVACEAE

Notes:
This plant is easy to grow in the home garden, where it is valued both for its beauty and its medicinal properties. To propagate, separate young offshoots from mature plants and transplant to your garden.

Saragundí

Geo-distribution:
This plant is native to the neotropics. In Costa Rica, it grows from low to high elevations (in fields, along roadsides, and in patios).

Botanical Description:
Saragundí is a bushlike perennial tree that grows to 2–4 m tall. The pinnate leaves consist of 6–12 pairs of leaflets. Racemes at the axils of the leaves bear bright yellow flowers.

Medicinal Uses:
A tea made from the leaves is used as a laxative and a diuretic. Fresh leaves are used to treat fungal infections of the skin, particularly ringworm (in Guatemala, the Afro-Caribbean name for this plant is ringworm bush). The flowers are said to be useful in the treatment of urinary problems. In Belize, women use a tincture to treat female infertility (Arvigo and Balick 1993).

Laxative, diuretic, vulnerary.

Preparation:
To make a laxative or diuretic, boil a handful of leaves in 1 L of water; drink several cups a day. To prepare a tea that treats urinary problems, boil a handful of flowers in 1 L of water. To treat skin infections, macerate fresh leaves and apply directly to the infected area. To make a tincture, soak a handful of finely chopped leaves in 250 mL of vodka (or clear rum) in a glass jar and seal the lid tightly. Shake the container once a day for a week before using.

Notes:
Saragundí is an easy bush to grow around the home; in addition to its medicinal uses, it serves as a lovely ornamental. Seeds can be collected from wild bushes.

SARAGUNDÍ
Senna reticulata
CAESALPINIOIDEAE

Sarsaparilla

ZARZAPARRILLA
Smilax medica, S. regelii,
and *S. ornata*
SMILACEAE

Geo-distribution:
Sarsaparilla is the name used to describe a group of native neotropical plant species that are found in rainforests from Mexico to Colombia.

Botanical Description:
Members of this group are thorny vines that start from an underground red tuber and then grow toward the forest canopy. They carry lanceolate to heart-shaped, alternative leaves. The flowers are green umbels with red or black seed capsules.

Medicinal Uses:
Indigenous tribes in many parts of the neotropics have used sarsaparilla to treat a variety of illnesses. It is reputed to have been successful in treating skin problems, including psoriasis and eczema. It is also said to be a blood purifier—effective in the treatment of anemia, energy depletion, acidosis, and a toxic state resulting from poor elimination (particularly in cases where poor elimination is due to kidney malfunction). Sarsaparilla tonic is said to stimulate the sex glands and act as an aphrodisiac. Traditionally, it has been used to treat impotence, liver problems, stress, rheumatism, gout, syphilis, leucorrhoea, herpes, epilepsy, fever, nervous disorders, diabetes, and stomach problems. Sarsaparilla is often referred to as "tropical ginseng."

Tonic, diuretic, vulnerary, depurative, febrifuge.

Preparation:
Boil 1 piece (2 cm^2) of the dried root in 1 L of water. Drink up to 3 cups per day.

Notes:
In Costa Rica, this plant is found only in primary rainforests; sarsaparilla does not grow well in home gardens. You can find commercially prepared sarsaparilla teas, however, in most macrobiotic and health food stores.

Scorpion's tail

Geo-distribution:
This herbaceous plant occurs in many neotropical countries. In Costa Rica, scorpion's tail (or *kĩ kuán* in the indigenous Boruca language) most commonly grows on the Pacific slope (from sea level to 1,200 m), where it is found in pastures, yards, clearings, and secondary-growth forests.

Botanical Description:
This annual plant, which has straight wood stems, grows from 50 cm to 2 m in height. It has several branches, each bearing rough leaves that are alternate and oval-lanceolate, with both ends pointed. The flowers are pastel blue and borne on a simple stem that resembles a scorpion's tail.

Medicinal Uses:
A maceration of fresh leaves is used to treat skin problems, particularly boils and abscesses. A decoction of the fresh leaves is used as a wash to treat eczema and athlete's foot. Scorpion's tail tea was used traditionally to relieve sore and inflamed throats, but internal use is no longer recommended (see the Caution).

Vulnerary.

Preparation:
Macerate fresh leaves and place on the affected area of the skin. To make a decoction, boil a handful of leaves in 1 L of water for about 10 minutes. Do not use internally (see the Caution).

Caution:
Research shows that scorpion's tail contains various carcinogenic compounds called pyrrolizidine alkaloids, which can pose serious health risks if taken internally. Continued internal use may cause liver damage or cancer; the plant should only be used externally (Arvigo and Balick 1993).

COLA DE ALACRÁN
Heliotropium indicum
BORAGINACEAE

Shrub verbena

LANTANA
Lantana camara
VERBENACEAE

Geo-distribution:
Shrub verbena—*lantana, soterrey, cinco negritos,* or *bandera española* in Spanish—occurs throughout Mesoamerica, the West Indies, and the southern United States. The plant grows in pastures and along roadsides in most parts of Costa Rica.

Botanical Description:
Shrub verbena has attractive yellow, orange, and red flowers that stand out against the greenery of the countryside; they attract butterflies and hummingbirds. The opposite, serrated leaves have a strong but pleasant aromatic odor, and are borne on square stems, a characteristic of the Verbenaceae family, to which this plant belongs. Most specimens grow to less than 1 m tall; however, in abandoned fields they often become tall bushes with prickly stems. The young flowers have four lobed petals that start out yellow then turn orange and red.

Medicinal Uses:
Note: This species is included in this book to document its traditional use as a medicinal plant. Today, however, scientists caution against its use; it is toxic.

Some Costa Ricans say shrub verbena can be used to treat a number of conditions (skin eruptions, for example), but researchers have discovered several toxic components in the plant. These toxins can cause skin irritation when used externally, and nausea, vomiting, and weakness when ingested—the plant is inappropriate for medicinal use (Robineau 2005).

Caution:
Do not use this plant for medicinal purposes—it is toxic.

Spiral flag

CAÑA AGRIA
Costus spicatus
and *C. ruber*
COSTACEAE

Geo-distribution:
Certain species of spiral flag are native to Central and South America. In Costa Rica, spiral flag plants grow in moist, shady areas from low- to midelevations.

Botanical Description:
Spiral flag is a succulent, gingerlike plant. It has large, lush leaves that are arranged in spirals around stems that rise from underground rootstocks. Bracted, conelike inflorescences appear at the tip of the stems.

Medicinal Uses:
The stems of these plants are used to make an aqueous solution for treating kidney ailments, skin disorders, and eye problems. *Diuretic, collyrium, antiherpetic.*

Preparation:
Chop a handful of fresh stems into small pieces and place in 1 L of water overnight. Drink the liquid in lieu of water throughout the day. Spiral flag is often mixed with coconut water or pineapple juice.

Notes:
These plants grow wild, but can also be cultivated in the home garden: divide the underground rhizome and then transplant to permanent sites.

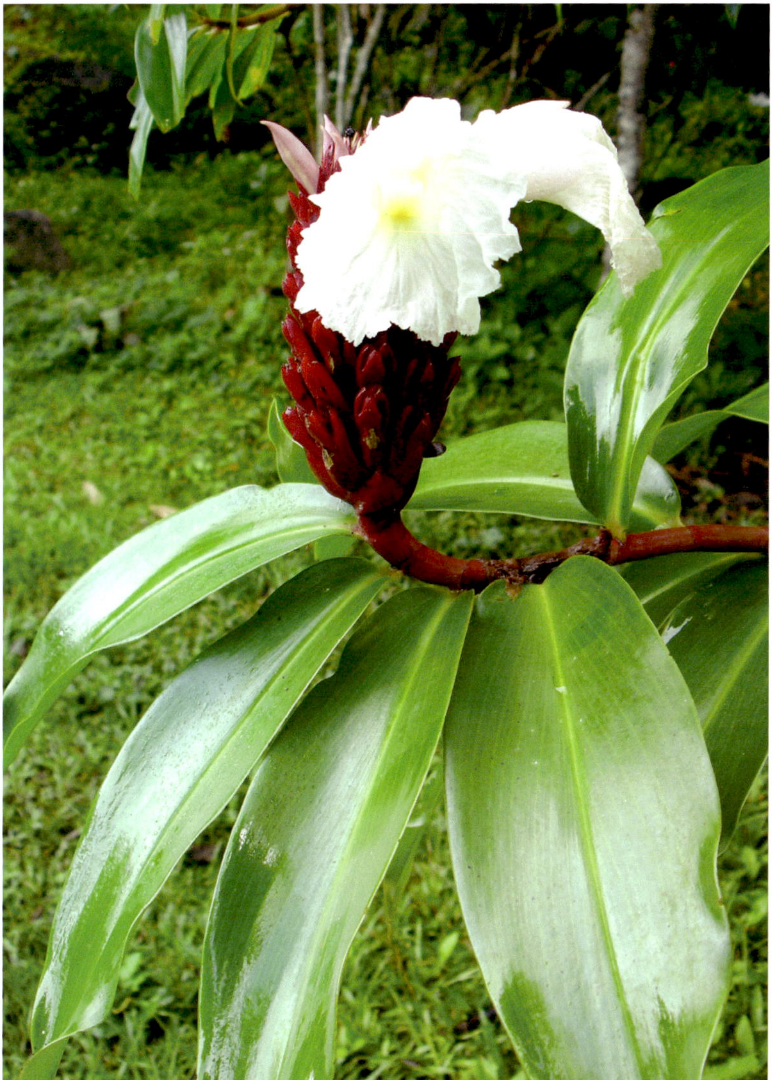

Stevia

Geo-distribution:
Native to Uruguay and Paraguay, *Stevia rebaudiana* is now grown in many parts of the world. It was introduced to Costa Rica in the 1990s and has been gaining popularity as a natural sweetener. It grows in commercial fields and home herb gardens.

ESTEVIA, HIERBA DULCE, or HOJA DULCE
Stevia rebaudiana
ASTERACEAE

Botanical Description:
Stevia is a perennial woody-stemmed herb with opposite, ovate, oblong, serrate leaves. Greenish-yellow flowers protrude in head-like clusters at the apex of the stem.

Medicinal Uses:
The benefits of stevia as a sweetener are unrivaled. This non-toxic plant, which belongs to the Aster family, contains stevioside, a crystalline diterpene glycoside that stimulates our sweet taste buds but does not contain sugars or calories. Stevia also has a number of traditional medicinal uses. Several indigenous tribes of South America have used it as a digestive aid; they also apply it topically to heal wounds. Recent clinical studies have shown that stevia increases glucose tolerance and decreases blood sugar levels; it is much safer than aspartame (Whitaker 1994).

Anti-diabetic, vulnerary.

Preparation:
To create a sugar-free sweetener, infuse the mature leaves in the herbal tea you want to sweeten. Dried leaves are more potent than fresh ones; one tablespoon of dried leaves is enough to sweeten 1 L of herbal tea.

Notes:
Although stevia is a relative newcomer to many parts of the tropics, it seems to adapt well to a wide variety of habitats at both lower and higher elevations. The author's trials show that it does extremely well as a potted plant in sunny areas around the home.

Stinging nettle

Geo-distribution:
Stinging nettle is commonly found at higher elevations in the tropics. (It is classified as a weed, and a noxious one at that—it provokes a rash when handled).

Botanical Description:
The stinging nettle is an herbaceous plant that grows to 50-100 cm tall; it has square, bristly stems and opposite, ovate, coarsely serrate leaves. The leaves and stems are covered with bristly hairs that inject an irritant under the skin when touched. The flowers appear in axial, oblong, dense, short spikes.

Medicinal Uses:
The stinging nettle is a nutritive herb with tonic effects; it has high concentrations of many mineral salts and vitamins. Traditionally, it has been used to stimulate digestion and increase the production of mothers' milk; it has also been used to relieve urinary tract infections, asthma, excessive menstrual flow, and anemia. Nettles are a good base for multiple-herb combinations.

Nutritive, tonic, expectorant, stomachic, emmenagogue, diuretic, pectoral, antianemic.

ORTIGA
Urtica dioica
URTICACEAE

Preparation:
Note: Steaming the nettles (or steeping them) neutralizes the stinging hairs.

Dried stinging nettles are available in health food stores and at some markets. Make an infusion by steeping 1 tablespoon of dried herb in 1 cup of water. You can also harvest young plants (pick plants that are no taller than 40 cm) for cooking (make sure to wear leather gloves!). Eating these steamed greens is a good way to boost the quantity of your mineral and vitamin intake.

Notes:
Stinging nettles are appropriate for the high-elevation tropical garden. Seeds are available from many herb companies.

Strangler fig

Geo-distribution:
In Costa Rica, the term *higuerón* refers to any member of the genus *Ficus*; larger species in this genus are also called *chilamates*. These trees are important pioneer and intermediate species in Costa Rica's tropical rainforests. They are found in most of the country's primary and secondary forests, principally near rivers; they are also planted in parks and town squares.

Botanical Description:
The strangler fig—known as *higuerón blanco* or *chilamate blanco* in Spanish—is a large tree that grows to a height of 25 m; it has an extensive canopy. The bright, shiny leaves are oblong and elliptical; they grow up to 25 cm long. The fruits are sometimes eaten by animals. The trunk excretes a white sap when an incision is made with a machete.

Medicinal Uses:
The white latex sap from the tree trunk's bark is used to treat parasitic intestinal worms, particularly whipworms (*Trichuris trichiura*).

Anthelmintic.

Preparation:
Make an incision in the tree trunk to collect the white sap. Mix 1 tablespoon of the latex in a glass of water; drink 1 glass before breakfast for 3 days.

Caution:
Although the two species mentioned here have been used for centuries as a vermifuge, many other species of this genus are toxic and may have serious secondary side effects if taken internally.

Notes:
These trees grow so large that they are not recommended for small home lots.

HIGUERÓN
Ficus glabrata and *F. jimenezii*
MORACEAE

Thyme

Geo-distribution:
Native to the Mediterranean region, thyme is now also found at higher elevations in the tropics. In Costa Rica, this small shrub is a popular plant for home patios.

Botanical Description:
This perennial, small (15–25 cm tall), bushy woody-stemmed herb has opposite sessile leaves that are ovate to lanceolate in shape and have slightly rolled edges. The small pastel blue–purple, two-lipped flowers are whorled in dense clusters.

Medicinal Uses:
Although thyme is best known as a seasoning, it also has medicinal applications. Thymol, the essential oil, helps treat colds and flu, sore throats, and coughs. It is also a strong antibacterial agent, which makes it useful in treating diarrhea, gastric infections, poor digestion, and gas. Using small amounts daily in salad dressings and other dishes helps prevent digestive problems.

Carminative, expectorant, pectoral, antibacterial, antidiarrheal, stomachic, anthelmintic.

Preparation:
Add several teaspoons of dried or fresh leaves to your salad dressing, or make your own dressing by adding thyme and other herbs, such as rosemary and oregano, to olive oil. To make an infusion, steep a handful of fresh thyme (or 1 tablespoon of dried thyme) in 1 L of water for 10 minutes. Drink no more than 1–2 cups per day.

Caution:
Like rosemary and oregano, thyme should be used with care. Overuse may cause secondary effects such as vomiting. Thyme oil is very toxic in its pure state; even small amounts can cause skin rashes.

Notes:
Thyme is easy to grow at home, either in pots or directly in the garden. Look for fresh cuttings in Costa Rican markets to start your own thyme at home.

TOMILLO
Thymus vulgaris
LABIATAE

Tradescantia

Geo-distribution:
Tradescantia (also called wandering jew or, in Spanish, *cucaracha*) is native to the neotropics; in Costa Rica, it occurs at lower elevations.

Botanical Description:
This plant is a creeping, herbaceous, perennial plant with alternate, variegated leaves of purple and silver tones. The small flowers are nestled in pastel purple leaf bracts.

Medicinal Uses:
Tradescantia has been used as an ornamental around the world, but few people are aware of its medicinal properties. The fresh juice is used to combat hemorrhages and neuralgia of the face.

Anodyne, antiherpetic, antiseptic, astringent, hemostatic.

Preparation:
Macerate the leaves to extract the juice; then apply the juice to bleeding cuts or painful areas.

Notes:
This is one of the easiest plants to grow in the home garden. Propagation is by vegetative cuttings, which can be planted in pots or directly in permanent sites. It also makes good ground cover in shady places.

HOJA DEL MILAGRO
Tradescantia zebrina
COMMELINACEAE

Tropical cedar

Geo-distribution:
Tropical cedar is native to the neotropics. It occurs at lower elevations, in secondary-growth forests and fields.

Botanical Description:
This tree grows to 20-30 m tall. The wood gives off a strong aromatic odor that is similar to cedar. The leaves are compound with 10-30 leaflets (each 7-13 cm long and 2.5-4.5 cm wide). Tropical cedar has small white flowers and 4 cm-long seed capsules (the seeds have winglike membranes).

Medicinal Uses:
A bitter tonic made from the inner bark treats infections (and colds and flu), purifies the blood, and strengthens the immune system; it also helps to remove phlegm and mucus from the lungs. Used externally as a wash, the tonic serves as a treatment for skin problems.

Tonic, antiseptic, expectorant, alterative.

Preparation:
The bark is prepared as a water extract, infusion, or decoction. Use one handful of inner bark per liter of water. Sip small amounts of the tonic throughout the day for no more than 3 days.

Caution:
Only use the tonic as an emergency botanical, since continued consumption may cause secondary effects such as nausea, vomiting, and dizziness.

Notes:
Tropical cedars are ideal timber trees in the tropics, but, because of their large size, they are not recommended for small home lots. To avoid leaving the trees vulnerable to disease, harvest bark only from the branches.

CEDRO
Cedrela mexicana
MELIACEAE

Tropical elm

Geo-distribution:
The tropical elm is a neotropical native that occurs at lower elevations.

Botanical Description:
This large tree grows to 20 m tall and has lanceolate-oblong, serrate, elmlike leaves. The flowers are white and have five petals; the mature fruits are black and round, with dense, pointed thornlike scales.

Medicinal Uses:
The inner bark of this tree is mucilaginous and is used to treat digestive problems, constipation, diarrhea, ulcers, and skin problems. A rinse made from the leaves is said to reduce hair loss.

Demulcent, stomachic, laxative, antiherpetic.

Preparation:
Prune a branch from the tree and remove the outer bark. Peel the cortex off down to the wood. Cut three 30-cm-long strips of cortex into 3 cm squares and soak in 1 L of water overnight. Drink the mucilaginous liquid throughout the day. To make a wash for the hair or skin, boil a handful of leaves in 1 L of water.

Notes:
Plant seeds in farm orchards. Livestock eat the mature, dried seedpods.

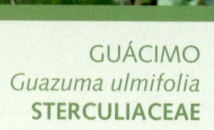

GUÁCIMO
Guazuma ulmifolia
STERCULIACEAE

Tuete

TUETE
Vernonia canescens H.B.K.
ASTERACEAE

Geo-distribution:
Tuete is distributed widely throughout the neotropics, including Central and South America. It grows at most elevations in Costa Rica, in secondary growth, pastures, lots, and along roads.

Botanical Description:
This plant is an herbaceous, bushlike perennial that grows to 2 m tall; it has numerous straight woody stems and gangly branches. The leaves are ovate to lanceolate and dentate. The small fragrant flowers are white or lavender.

Medicinal Uses:
Tuete is used primarily as a wash to remedy skin problems (such as cuts and bruises), to stop bleeding, and to reduce swelling. Fresh and dried flowers are used to treat constipation.

Vulnerary, laxative.

Preparation:
In Costa Rica, tuete has a long history as an herbal first aid. Campesinos have traditionally used it to treat wounds and bruises. The fresh leaves are mashed or chewed and applied directly to the affected area of the skin. An infusion of a handful of fresh leaves to 1 L of water is used to bathe the skin. To treat constipation, infuse a handful of fresh flowers or a tablespoon of dried flowers in 1 L of water; drink 1–2 cups a day.

Notes:
No toxicity levels have been reported for tuete. Since leaves and flowers are readily harvested in the wild, this plant is not a must-have for the home garden.

Turmeric

CÚRCUMA
Curcuma longa
ZINGIBERACEAE

Geo-distribution:
Turmeric was originally cultivated in the Old World tropics, but it now thrives in other tropical regions around the world.

Botanical Description:
This is an herbaceous perennial plant of the ginger family with a bright orange 10–15 cm-long rhizome. The central stem has large, dark-green leaves that are either obliquely erect or oblong and lanceolate. These leaves taper near the leaf apex, broaden near the base, and envelop the succeeding shoot. The yellowish white flowers are borne on cylindrical spikes that carry numerous greenish white bracts. During the dry season, the stem and leaves dry up and the rhizome is dormant.

Medicinal Uses:
Turmeric has a long history of being used for both food and medicine. Ancient Indo-European Aryan cultures, who worshiped the sun, attributed to turmeric the pro-

tective powers of sunlight. (In Sanskrit, turmeric is known as *haridra*). It has been a staple in Ayurvedic medicine for 6,000 years; Ayurvedic practitioners have used it to purify the blood and to treat a number of illnesses. Capsules of ground turmeric powder have been used to relieve stomach problems, indigestion, flatulence, liver and gallbladder diseases, arthritis and rheumatism, catarrh, and colds and flu. Raw turmeric applied to the skin is effective in treating inflammations, infections, bruises, and sprains.

Since the discovery of turmeric's antioxidant phenolic compounds—and the protection that these compounds provide against free radicals—this spice is now viewed as much more than just an ingredient in curry or yellow dye. Turmeric's potential use in cancer prevention and the treatment of HIV infections is now the subject of intense laboratory and clinical research. For a complete explanation of turmeric's unique biochemistry and a survey of recent research, see *Turmeric and the Healing Curcuminoids* (Majeed, Badmaev, and Murray 1996).

Stomachic, hepatic, antibacterial, anti-inflammatory, antiviral, cholagogue, depurative, hepatic, vulnerary.

Preparation:

Using lots of turmeric in your food is a good way to prevent health problems. A small quantity of raw, grated turmeric adds taste to salad dressings and gives them a bright orange color. (Curries also contain a lot of turmeric).

Take 3-6 capsules of ground turmeric powder per day. To make an external liniment, combine 1 aloe vera leaf, 1 turmeric root, and 1 ginger root in a blender (do not add water); the resulting gelatin can be used as a poultice for joint or lower-back pain, sprains, and several kinds of skin problems. To make a tincture, use the same ingredients as for the liniment: 1 aloe vera leaf, 1 turmeric root, and 1 ginger root. Chop the ingredients into small pieces and add them to 1 L of vodka. Let the solution sit for a week before using it.

Notes:

Turmeric is easy to grow in your home garden, which will produce sufficient quantities of roots for salad dressings, homemade curries, and medicinal use. (In Costa Rica, you can frequently find turmeric roots in supermarkets and at herb stands.)

Vervain

Geo-distribution:
Originally from Europe, vervain was later introduced to the neotropics. In Costa Rica, it grows in fields and near homes, both at midelevations and in the highlands.

Botanical Description:
This herb consists of a whitish, branched, spindle-shaped root or rootstock; it produces a stiff, quadrangular stem that branches near the top. The leaves are opposite, oblong to lanceolate, entire and sessile at the top, and deeply cleft at the bottom. The flowers are white to pastel purple on slender spikes.

Medicinal Uses:
This little plant has remarkable healing powers. It is well-known as a treatment for colds and flu, fevers, nervous disorders, kidney problems, liver problems, and jaundice; it is also used as a wash for skin problems.

Astringent, diuretic, emmenagogue, stimulant, tonic, vulnerary, nervine, sudorific, galactagogue.

Preparation:
To make an infusion, steep a handful of fresh leaves in 1 L of water for 10–20 minutes. Drink 1–2 cups per day until results are obtained. The plant has a strong taste; mix it with other, more flavorful, herbs to mask the bitterness.

Notes:
This medicinal herb thrives in home gardens at higher elevations in the tropics. Many herb and seed companies sell vervain, and fresh cuttings are sometimes found in Costa Rican markets and herb stands.

VERBENA
Verbena litorales
VERBENACEAE

Wild sage

SALVIA VIRGEN
Buddleja americana
LOGANIACEAE

Geo-distribution:
Wild sage is native to Mexico and Central America. In Costa Rica, it is usually found from mid- to high elevations, in the Central Valley and on the Pacific slope. It occurs in secondary-growth forests and along fences and pastures, but it is also planted in patios for medicinal use.

Botanical Description:
This plant is a perennial bush that grows 1-2 m tall. The rectangular stems bear leaves that are opposite, elliptical, oblong, hairy, and serrated; the underside of each leaf is grayish green. The leaves give off a camphorlike odor. The small pale yellow flowers are borne on panicles.

Medicinal Uses:
When applied as a maceration, wild sage's camphorlike resins help heal lacerations, bruises, strains and sprains. Externally applied infusions are used to treat rashes and skin infections; taken internally, infusions have reportedly been successful in relieving insomnia symptoms, menstrual problems, hemorrhoids, digestive disorders, nervous conditions, headaches, upper respiratory infections, and asthma.

Antiherpetic, relaxant, emmenagogue, stomachic, nervine, expectorant, pectoral.

Preparation:

To prepare a maceration, simply mash a handful of the leaves; apply to the affected area. To make an infusion, steep 2–6 leaves per cup of boiling water for 10–15 minutes. Drink 1–3 cups of tea per day.

Notes:

This popular tropical medicinal plant is often sold at small Costa Rican markets and herb stands. Fresh, 15–30 cm-long woody stem cuttings can be propagated at home in plastic nursery bags or in pots filled with fertile soil. Keep the cuttings in a shady place and water them frequently until they begin to sprout new leaves; then, incrementally, move the cuttings toward a sunlit area, and water 2-3 times per week. Once the plants are well established, they can be transplanted to a permanent site in the herb garden. This plant grows best when planted in well-drained soil that receives full sun. Because of its large size when mature, it is best placed along the edges of the garden. Its beautiful silver-gray leaves make it an attractive ornamental.

Wild tobacco tree

Geo-distribution:
Wild tobacco grows at most elevations in Costa Rica (up to 2,000 m). It occurs naturally in wet, humid areas of secondary mountainous growth; because wild tabaco makes an ideal host tree for orchids, it is planted in backyards and patios throughout the country.

Botanical Description:
Wild tobacco grows to 3–5 m tall with alternate, peculate, glabrous leaves. These leaves are elliptical, oblong, and pointed at both ends. The unique beige bark is deeply furrowed and has a corklike appearance.

Medicinal Uses:
Juice extracted from young shoots and tea made from the leaves are both used externally to soothe hemorrhoid pain. The leaves have also been used to treat dandruff and relieve sore throats (see Caution).

Vulnerary.

Preparation:
Macerate young stems to extract the juice. To make a bath that is said to relieve hemorrhoid pain, boil a handful of leaves in 1 L of water. To make a hair rinse to treat dandruff, soak a handful of leaves overnight in 1 L of water. The young leaves were formerly used to make a gargle to soothe sore and inflamed throats (see Caution).

Caution:
The leaves of this tree are toxic and you should therefore never drink teas that are made from them (Vargas Chinchilla 1990).

Notes:
Wild tobacco can be propagated from branches that are planted directly in the soil. These trees are also popular among orchid growers, because orchid roots can easily attach themselves to the soft, corklike bark of the trunk and branches (without harming the tree). The tree also provides support and shade for the orchids.

GÜITITE
Acnistus arborescens
SOLANACEAE

Medicinal Plants of Costa Rica / **111**

Winged-leaved quassia

Geo-distribution:
The winged-leaved quassia tree is native to the neotropics, where it is particularly common in the Greater Antilles and Central America. In Costa Rica, it generally occurs at low- to midelevations, in fields and in primary and secondary forests.

Botanical Description:
This is a slender tree with brownish gray bark. The light green, pinnate, compound, glaucous leaves have prominent veins on the lower surface; the leaflets are obtuse or rounded at the apex. The ovoid to oblong-ovoid fruits (1.5–2 cm long) are red at first, then turn black.

Medicinal Uses:
The root, bark, and wood, which are all low in toxicity, are used to make a safe emergency tonic for poor digestion. The tree is sometimes called dysentery bark, because it is used to treat intestinal parasites, diarrheas, and fevers.

Stomachic, antiparasitic, febrifuge, antidiarrheal.

Preparation:
To make an infusion or decoction, soak 2 to 3 segments (2.5 cm) of the bark or root in 1 L of water. Recommendations call for drinking 3 cups per day, although the strong, bitter taste of the liquid makes it unpleasant to swallow.

Notes:
The bark and roots are sold at herb stands in Costa Rica. Seedling trees can be planted in tropical home gardens at most elevations (from sea level to 1,500 m).

ACEITUNO NEGRO
Simarouba glauca
SIMAROUBACEAE

Wormseed

APAZOTE
Chenopodium ambrosioides
CHENOPODIACEAE

Geo-distribution:
Wormseed grows in most parts of the neotropics. In Costa Rica, it is occasionally planted in home patios for medicinal use.

Botanical Description:
This herbaceous plant has a strong, distinctive odor, somewhat like that of camphor. Wormseed grows to about 1 m tall, with many branches and small leaves. These leaves are alternate, dentate, and ovate-lanceolate in shape. The inconspicuous green flowers are formed on dense spikes.

Medicinal Uses:
Throughout the tropics, wormseed is a popular antiparasitic medicine. It contains ascaridol, a compound that exerts a paralyzing, narcotic effect on Ascaris and Ancylostoma worms (although it is not strong enough to combat either tapeworms or Trichocephalus worms). According to the World Health Organization, a single 20 g dose rapidly expulses the target parasites with no secondary effects (Robineau 2005).

Antiparasitic, anthelmintic.

Preparation:
Boil 30 g of the fresh plant in 500 mL of milk. Take 3 times a day before meals (for no more than 6 days); during this treatment, castor oil is often also administered to flush out parasites.

Caution:
Overuse may cause vomiting and other secondary effects. Pregnant women and people with low energy should not take this treatment.

Yarrow

Geo-distribution:
European colonists introduced yarrow to the neotropics, where today it thrives at higher elevations.

Botanical Description:
This herb has finely dissected dentate segments; the leaves are narrowly oblong, lanceolate, alternate, and bipinnatifid. Its stem bears grayish white or rose-colored flowers on compound corymbs.

Medicinal Uses:
Yarrow has been used as a medicinal plant in Asia and Europe for a very long time. Its traditional medicinal uses include treatment for fevers, flu, colds, diarrhea, uterine problems, headaches, hair loss, and eruptive diseases such as measles and chicken pox. As a tonic tea, it helps boost the body's natural defenses, and it is often given to those recuperating from illness or surgery.

Diaphoretic, diuretic, stimulant, astringent, tonic, alterative, emmenagogue, vulnerary.

Preparation:
To make an infusion, steep a handful of fresh leaves or 1 teaspoon of dried leaves in 1 L of water for 15 minutes. Let the infusion cool before using it as a bath, skin wash, douche, or enema.

Notes:
Yarrow thrives at higher elevations in the tropics; it is propagated by seeds.

MILENRAMA
Achillea millefollium
ASTERACEAE

Yellow dock

ROMAZA or RUIBARBILLO
Rumex crispus
POLYGONACEAE

Geo-distribution:
Yellow dock grows in temperate climate zones around the world. Perhaps an Old World native, this plant was reportedly introduced to the New World by early settlers—the animals they brought with them defecated out the seeds. In the neotropics, it occurs in the cooler highlands, growing in pastures and other areas with wet soil.

Botanical Description:
Yellow dock is a perennial herbaceous plant that has large (15-35 cm-long) green leaves with a lancelike appearance and a curly border. The stems are striped with a red center. The small, green flowers, which turn into dried, red seed capsules, are borne on an erect stem. The 10-20 cm-long root is bright yellow on the inside. Yellow dock is often confused with rhubarb (*R. rhaponticum*).

Medicinal Uses:
European settlers have used yellow dock for centuries to treat a variety of health problems. The root has strong, bitter, yellow alkaloids, much like those in goldenseal (*Hydrastis canadensis*), and can be used to treat skin problems such as psoriasis, eczema, and urticaria. Yellow dock has been taken internally to remedy iron deficiencies and anemia, hemorrhoids, bile congestion, rheumatism, infections, and digestive disorders; it is also taken internally for its laxative and general tonic effects. This is an emergency botanical and should not be taken for any longer then seven days at a time.

Astringent, antiseptic, antibacterial, hepatic, digestive, laxative, tonic.

Preparation:

Use the fresh or dried mature root of the plant to make teas or tinctures. To prepare dried root, first wash the roots thoroughly and then slice them into thin sections. Dry the sections either in a warm place or in the oven until they are brittle; store them for future use. To make a tea, follow these instructions:

1. Put the dried sections in a blender and grind to a powder.
2. Add 1 teaspoon of the root powder (or several slices of fresh root) to each cup of water.
3. Boil for 10-20 min.
4. Drink 1-3 cups of tea per day for 3-6 days.

To make a tincture, add the dried or fresh root to vodka or clear rum. Shake daily for several weeks before using. Take 1-3 teaspoons per day for 6 days. You can also use this tincture for extended periods to treat skin externally.

Although the root is more potent medicinally than any other part of the plant, you can add the tender young leaves to salads.

Notes:

Yellow dock root can be used for medicinal purposes in lieu of goldenseal (*Hydrastis canadensis*), an endangered species that grows in temperate forests. Gardeners in the tropical highlands will find that this plant makes a nice addition to their garden collection. You can either transplant young plants from pastures to the home garden or collect seeds and sow them in a germination unit. Yellow dock is an extremely hardy plant and requires no special care. Make sure, however, to plant in fertile soil, in an area with moist to wet conditions; the planting site should receive partial to full sun.

Yucca

Geo-distribution:
Native to the neotropics, yucca grows at most elevations in Central and South America, Mexico, and the southern United States. People in Costa Rica commonly plant yucca to make living fences around fields and homes.

Botanical Description:
This plant grows to 3-4 m tall with a long, columnlike trunk that has few or no branches. 70-cm-long, erect lanceolate leaves with spiked tips appear at the apex of the trunk. The bell-like flowers are cream-white and about 4 cm long; they form pastel yellow-green, oblong, ovoid fruits that grow to 2-3 cm long.

Medicinal Uses:
The beautiful flowers of this plant have medicinal properties and are also a tasty ingredient in several Costa Rican dishes. They contain bitter constituents said to be useful as a stomach tonic. In March and April, especially during Easter Week, Costa Ricans use the flowers to make a traditional *picadillo* dish that includes eggs; the flowers can also be added to salads. The leaves are employed as a diuretic and for the treatment of colitis. Research has shown that the stems contain saponins, which are used in the treatment of arthritis and rheumatism. Commercial medicinal products made from this plant are now available in the United States.

Stomachic, diuretic, antiarthritic.

Preparation:
Sauté the flowers with eggs and onions to prepare the popular Easter dish, or steam and serve in salads. Boil a handful of flowers in 1 L of water to prepare a tea that is diuretic and that helps treat albuminuria (indicated by the presence of albumen in the urine). To prepare a tea using the leaves, boil five 10 cm-long pieces in 1 L of water for 3 minutes. Drink 3 cups per day.

Notes:
The hardy yucca is an ideal home garden plant; it requires no special care or watering, even during the dry season. To propagate this plant, use the stem cuttings (5-10 cm in diameter and 1-2 m tall); plant them 30 cm deep in permanent sites. In Costa Rica, it is best to plant yucca in April or May; this gives the plants the entire rainy season to produce roots before the dry months arrive. The flowers appear each year during the dry season.

ITABO
Yucca guatemalensis
AGAVACEAE

Medicinal Plants of Costa Rica / **117**

Zornia

Geo-distribution:
Zornia is native to Central America; it is common at most elevations in Costa Rica, particularly on the Pacific slope and in the Central Valley. It grows along roads, in open fields and vacant lots, and in forest areas with partial shade. Zornia is frequently considered invasive in pastures and at home sites.

ZORNIA or TRENCILLA
Blechum pyramidatum
ACANTHACEAE

Botanical Description:
Zornia is a perennial herbaceous plant that grows to 30 cm tall; it has square stems that stand upright. The plant has aromatic, pubescent, crenate, opposite leaves. The flowers are formed on spikes with leaflets that resemble a French braid, hence the Spanish name *trencilla* (braid).

Medicinal Uses:
Traditionally, zornia has been used to make a tea to relieve stomachaches and to combat amoebas and other micro-parasites that produce diarrhea and fever. This tea, which has a mild flavor, is also suitable for treating children.

Stomach tonic, antiparasitic, febrifuge, antidiarrheal.

Preparation:
Zornia is administered as a concentration. Boil a handful of leaves, stems, and flowers in 1 L of water until the volume of water is reduced to 500 mL. Drink small cups of this concentrate throughout the day. In acute cases of dysentery, combine zornia with winged-leaved quassia (*Simarouba glauca*) to increase its effectiveness; use 3 pieces of winged-leaved quassia bark (2.5 cm long) per liter of water. In mild cases of diarrhea, steep a handful of fresh leaves in 1 L of water for 20 minutes to make a mintlike tea; add honey to sweeten the drink.

Notes:
Because zornia grows abundantly at most elevations in Costa Rica, it is easy to collect in the wild. To propagate zornia, dig up wild plants and transplant them to the herb garden. This plant grows best in moist soil with partial shade.

Making Herbal Preparations

Macerations and water extracts with fresh plants

The use of macerations is perhaps the simplest—and often the best—way to apply tropical medicinal plants for treating skin problems. It's easy to create aloe macerations; just snap a leaf in two and apply the resulting gel to the skin—or even the eyes when they are sore or disturbed by foreign particles. With other plants, however, you will need to hand crush them or macerate them in a mortar in order to release their juices. Alternatively, many types of plant material can be macerated by running the material through a juicer. Extracted plant juices are concentrated and usually should be diluted if taken internally; as a general rule, mix 1 tablespoon to 1/2 cup of water.

Water extracts are another effective way of preparing treatments for external and internal health problems. The procedure is simple: First cut the plant material into small pieces and add water. In general, use about one handful of chopped plant material per liter of water; let stand overnight. The following day, drink the solution in place of your daily water intake. Many original indigenous remedies use water extracts as a tea or bath.

Sun tea

In the tropics, you can use the power of the sun to make herbal teas, thus saving energy and preserving vital organic compounds. Just put one handful of fresh herbs with one liter of water in a clear glass jar. Cover the jar tightly and place it in strong morning sunlight. If you put out the jar in the early morning, the tea should be ready by the afternoon.

Infusions

In simple terms, this refers to the process of steeping herbs in hot, but not boiling, water—ideally in a ceramic teapot. This method is preferred for aromatic leafy plants that would lose much of their essential oil if boiled. Dried aromatic herbs make a stronger-tasting infusion than fresh herbs. Let the boiling water cool slightly before adding it to the teapot.

Decoctions or concoctions

Boiling extracts organic compounds from certain plant material, particularly stems, roots, and barks. How long you should boil plant material depends a great deal on its form. For example, powdered roots and barks are boiled for only a short time, while larger pieces take longer. Gumbo limbo, bitterwood, and tropical cedar are just a few of many tropical barks that should be boiled to best bring out their medicinal properties. Ginger root should always be steeped rather than boiled, since some of its aromatic oils can be lost in vapors.

Concentrations

It is often useful to concentrate a decoction by eliminating the excess water. In this process, the solution is boiled until half the liquid is gone. The final concentration can be preserved by adding alcohol (vodka or clear rum) or honey. Many cough syrups are made this way.

Syrups

One of the trickiest parts of treating sick children is getting them to take herbal medications. Syrups are an age-old solution. Here's a tropical cough syrup that helps children of all ages: Boil 50 g pink trumpet tree or pau d'arco bark, 50 g turmeric root, and 50 g ginger root in a liter of water. Concentrate to 250 mL and strain the liquid. Add 250 mL of pure honey to the liquid. Store in a dark bottle. Suggested dosage: 3-6 teaspoons per day.

Alcohol extracts and tinctures

Alcohol extracts and tinctures are convenient ways of preserving herbal medicine and are easy to prepare in your kitchen. An extract has less than 50 percent alcohol; a tincture has 50 percent or more. Homemade extracts or tinctures are usually made from a base of vodka, which has a relatively high proportion of alcohol to water, allowing both water- and alcohol-soluble compounds to dissolve well.

If you use fresh plant material, half-fill a glass container with herbal material; add enough vodka to top off the container. Use a wide-mouthed gallon or liter jar.

Powdered herbs can also be preserved in the same way, though in this case you should use less herbal material and more vodka. Use 60–100 g of powder per liter of vodka.

Oil extracts

Oil is another base that you can use to preserve herbs and other plants, and oil extracts are easy to prepare in your home kitchen. (Wide-mouthed jars work well for these extracts.) To make a classic recipe from Asia: Half-fill a container with grated ginger and top off with sesame oil. Let the mixture stand for several days, shaking it frequently; strain and store in a dark bottle. This oil extract is used externally for sore muscles, strains, and sprains.

Capsules

Empty gelatin capsules are now available at most health food stores; they are a useful way of administering herbal medicines. Using a blender, grind herbal mixtures to a fine powder. Next, open the capsules and fill them with the powder. Store both empty and filled capsules in air-tight containers to prevent moisture damage. Dosages range from 3 to 6 capsules per day, depending on the ingredients.

Compresses and poultices

A poultice is a mixture of various herbs wrapped in material, such as cheese cloth, and applied to the skin. Poultices generally relieve sore muscles and joints or stimulate internal organs, such as the liver or lungs. Equal parts of freshly blended ginger, turmeric, and aloe vera make a good poultice for the treatment of many ailments. Apply externally for 30 minutes or more. Heating the poultice increases effectiveness.

A compress is a hot herbal tea preparation that, applied to the body with a towel, soothes sore muscles and joints or, applied to the chest, helps treat colds.

Conversion chart

Liters to cups	1 L = 4.2 cups
Milliliters to cups	250 mL = 1 cup
Kilograms to pounds	1 kg = 2.2 pounds
Grams to ounces	1 g = 0.035 ounce
Centimeters to inches	1 cm = 0.4 inch
Meters to feet	1 m = 3.3 ft

Matera Medica

An alphabetical index of common health problems with recommendations for plants that may aid in the treatment of those conditions.

Acidosis: aloe vera, China root, sarsaparilla

Anemia: China root, dandelion, guapinol, pink shower tree, pink trumpet tree, sarsaparilla, stinging nettle, yellow dock

Arthritis and rheumatism: aloe vera, avocado, cecropia tree, dandelion, echinacea, feverfew, garlic, gotu kola, hoja de estrella, horsetail, juanilama, noni, parsley, pink trumpet tree, pisabed, prickly pear cactus, sarsaparilla, turmeric, yellow dock, yucca

Asthma: cecropia tree, eucalyptus, garlic, parsley, pink trumpet tree, rose periwinkle, stinging nettle, wild sage

Backache: ginger, pink shower tree, turmeric

Bacterial infections: aloe vera, cayenne, echinacea, garlic, ginger, Indian almond, jackass bitters, mint, pink trumpet tree, purple mombin, red targua, redhead, sansevieria, peppermint and spearmint, thyme, turmeric, yellow dock

Boils: dandelion, garlic, pink trumpet tree, rat killer tree, scorpion's tail

Cancer: aloe vera, dandelion, garlic, pink trumpet tree, rose periwinkle, turmeric

Colds and flu: avocado, borage, broom weed, cayenne, cecropia tree, citrus, echinacea, elderberry, eucalyptus, feverfew, garlic, ginger, gotu kola, greater plantain, gumbo limbo, juanilama, lemongrass, life everlasting, mozote de caballo, oregano, pink trumpet tree, pisabed, thyme, tropical cedar, turmeric, vervain, yarrow

Colitis: aloe vera, pink trumpet tree, yucca

Constipation: aloe vera, castor bean, chan, dandelion, dwarf poinciana, golden shower tree, papaya, pink shower tree, pink trumpet tree, rose periwinkle, tropical elm, tuete

Coughs: avocado, broom weed, echinacea, elderberry, eucalyptus, greater plantain, oregano, parsley, pink trumpet tree, peppermint and spearmint, thyme

Cramps: basil, dandelion, pisabed, peppermint and spearmint

Depression: passion flower, dwarf poinciana

Diabetes: bitter gourd, dandelion, guapinol, noni, pink trumpet tree, rose periwinkle, sarsaparilla

Diarrhea: annatto, avocado, bitterwood, cilantro, cinnamon, coconut, guapinol, guava, gumbo limbo, Indian almond, jackass bitters, mozote de caballo, pink trumpet tree, pisabed, purple mombin, rat killer tree, thyme, tropical elm, winged-leaved quassia, yarrow, zornia

Digestive problems: allspice, aloe vera, basil, cayenne, chamomile, chan, cinnamon, citrus, coconut, feverfew, ginger, juanilama, lemongrass, mint, mozote de caballo, oregano, papaya, parsley, pink shower tree, prickly pear cactus, rosemary, sage, peppermint and spearmint, stevia, stinging nettle, thyme, tropical elm, turmeric, vervain, wild sage, winged-leaved quassia, yellow dock

Fever: annatto, artemisia, avocado, bitterwood, borage, broom weed, castor bean, cecropia tree, dandelion, echinacea, elderberry, eucalyptus, feverfew, gumbo limbo, jackass bitters, lemongrass, pink trumpet tree, pisabed, purple mombin, sage, sarsaparilla, vervain, winged-leaved quassia, yarrow, zornia

Fungus: garlic, jackass bitters, pink shower tree, redhead, sanseviera

Gallbladder: aloe vera, pink trumpet tree, turmeric

Head lice: jackass bitters, noni, pokeweed

Hair loss: aloe vera, annatto, bluebush, coconut, elderberry, horsetail, rosemary, tropical elm, yarrow

Headaches and migraines: annatto, avocado, citrus, feverfew, gumbo limbo, hoja de estrella, life everlasting, oregano, passion flower, redhead, peppermint and spearmint, wild sage, yarrow

Heart and circulatory problems: cayenne, coconut, garlic, ginger, gotu kola

Hemorrhoids: Indian almond, wild sage, wild tobacco tree, yellow dock

Inflammations: aloe vera, castor bean, cecropia tree, dandelion, lemongrass, pisabed, turmeric

Insect bites and stings: aloe vera, rat killer tree, redhead

Insomnia: dandelion, garlic, gotu kola, mimosa, mint, passion flower, peppermint and spearmint, wild sage

Jaundice: avocado, dandelion, parsley, pisabed, purple mombin, vervain

Kidney and bladder disorders: aloe vera, annatto, cayenne, China root, coconut, cornsilk, dandelion, ginger, gumbo limbo, horsetail, mozote de caballo, papaya, parsley, pink shower tree, sarsaparilla, spiral flag, vervain

Liver problems: aloe vera, bitter gourd, cayenne, citrus, coconut, dandelion, garlic, greater plantain, juanilama, papaya, pink trumpet tree, rosemary, sarsaparilla, turmeric, vervain

Measles: elderberry, gotu kola, yarrow

Motion sickness: ginger

Mumps: elderberry

Nervousness: basil, carpenter's bush, chamomile, citrus, gotu kola, juanilama, lemongrass, mimosa, mint, oregano, passion flower, sage, sarsaparilla, peppermint and spearmint, vervain, wild sage

Pancreas problems: bitter gourd, dandelion

Senility: dandelion

Skin problems: aloe vera, avocado, bitter gourd, broom weed, cayenne, China root, cinnamon, coconut, dandelion, elderberry, eucalyptus, gotu kola, greater plantain, guava, gumbo limbo, hibiscus, hoja de estrella, horsetail, Indian almond, jackass bitters, life everlasting, noni, oregano, papaya, pink shower tree, pink trumpet tree, pisabed, prickly pear cactus, purple mombin, rat killer tree, red targua, redhead, rosemary, sansevieria, saragundí, saragundí, sarsaparilla, scorpion's tail, shrub verbena, spiral flag, tropical cedar, tropical elm, tuete, turmeric, vervain, wild sage, yarrow, yellow dock

Sore throat: cecropia tree, ginger, gotu kola, greater plantain, pink trumpet tree, rose periwinkle, scorpion's tail, thyme, wild tobacco tree

Sore muscles: arnica, ginger, hoja de estrella, rosemary

Strains and sprains: aloe vera, arnica, ginger, life everlasting, turmeric, wild sage

Stress and anxiety: carpenter's bush, citrus, gotu kola, lemongrass, mimosa, oregano, passion flower, sage, sarsaparilla, vervain, wild sage

Tonsillitis: dandelion, gotu kola

Ulcers: aloe vera, coconut, garlic, mozote de caballo, purple mombin, tropical elm

Warts: garlic, papaya

Worms: avocado, bitter gourd, castor bean, coconut, feverfew, garlic, jackass bitters, parsley, passion flower, pink shower tree, saragundí, strangler fig, wormseed

Glossary

bot.= botanical term

abortive: An agent that induces abortion.
abscess: A puss-filled cavity, surrounded by swollen, inflamed tissues.
acidosis: Depletion of the body's alkali reserve, with resulting disturbance of the acid-base balance in the body.
acuminate: *bot.* Tapering gradually to a point at the apex.
acute: *bot.* Sharp, ending in a point.
albuminuria: The presence of albumen protein in the urine.
alterative: An agent capable of favorably altering or changing unhealthy conditions of the body; tends to restore normal bodily function.
alternate: *bot.* Arranged first on one side of a stem, then on other side further along the stem, or at different points along the stem or axis.
analgesic: A medication which relieves or diminishes pain (see anodyne).
anemia: A condition in which there is a reduction of the number of red blood cells and hemoglobin in the body.
anesthetic: An agent that deadens sensation.
annual: *bot.* Completing the cycle from seed to death in one season.
anodyne: An agent that allays or kills pain.
anthelmintic: An agent used to eliminate or destroy parasitic intestinal worms.
antibiotic: An agent that destroys or arrests the growth of microorganisms.
antidiarrheal: An agent that is used to treat diarrhea.
antiherpetic: An agent useful against skin diseases.
antiparasitic: An agent that eliminates parasites.
antiseptic: An agent that inhibits the growth of micro-organisms on living tissue.
antispasmodic: An agent used to ease muscular spasms or convulsions.
aperient: A mild stimulant for the bowels; a gentle purgative.
aromatic: A plant or medicine with a fragrant smell and often a warm pungent taste.
arteriosclerosis: A thickening or hardening of the artery walls.
astringent: A substance that causes contraction of tissues, checking the discharge of mucus and fluid from the body.
axially: *bot.* Developing from the main axis or main stem.
biennial: *bot.* Completing the cycle from seed to death in two years or seasons.
bipinnate: *bot.* Having 2 sets of leaflets arranged in opposite rows along the petiole.
bile: The bitter fluid secreted by the liver into the gall bladder; it aids in digestion.
bipinnatifid: *bot.* Having 2 sets of cleaved leaflets arranged in opposite rows along the petiole.
bitter: *bot.* Characterized by a bitter substance that acts on the mucous membranes of the mouth and stomach to increase appetite and promote digestion.
bract: *bot.* A small, sometimes scale-like leaf, usually associated with flower clusters.
calmative: An agent that has a mild sedative or tranquilizing effect.
calyx: *bot.* The outer part of a flower, usually consisting of green, leafy sepals.
cambium: *bot.* The soft cellular tissue layer—between the bark and the wood—from which new bark and wood arise. It is composed of phloem and xylem cells.

carminative: An agent used to relieve colic, griping, or flatulence, or to expel gas from the intestine.

catarrh: An inflammation of a mucous membrane (usually the nasal and air passages) characterized by congestion and the secretions of mucus.

cathartic: An agent used to encourage the evacuation of the bowel (a laxative or purgative).

cholagogue: An agent that increases the flow of bile into the intestines.

colic: Paroxysmal pain in the abdomen or bowel due to various abnormal conditions in the colon.

colitis: Inflammation of the mucous membrane of the large intestine.

collyrium: Any medicated preparation for the eyes; eyewash.

compress: A hot solution of herbal tea applied with a towel to the skin to aid in healing.

corymb: *bot.* A type of inflorescence that resembles a raceme and that forms a flat-topped or convex cluster.

corolla: *bot.* The petals of a flower, which may be separate or joined in varying degrees.

crenate: *bot.* Having rounded teeth along the margin.

cyme: *bot.* A branching, relatively flat-topped flower cluster whose central or terminal flower opens first, forcing development of further flowers from lateral buds.

deciduous: *bot.* Annual shedding of leaves.

decoction: A preparation made by boiling a solution of plant parts and some kind of liquid (often water).

demulcent: A medicinal liquid of a mucilaginous nature taken internally to soothe inflamed mucous surfaces and to protect them from irritation.

dentate: *bot.* Sharply toothed, with the teeth pointing straight out from the margin.

depurative: An agent that cleanses and purifies the body, particularly the blood.

diaphoretic: An agent that promotes perspiration; sudorific.

digestive: An agent that promotes or aids digestion.

diuretic: An agent that increases the secretion and expulsion of urine.

dropsy: An abnormal accumulation of fluid in the body.

drupe: A fruit that has leathery skin, soft pulp, and a stony pit.

dyspepsia: A condition of disturbed digestion characterized by nausea, heartburn, and gas.

eczema: A disease of the skin characterized by itching, inflammation, and the formation of scales.

emmenagogue: An agent that stimulates menstrual flow.

emollient: A substance of mucilaginous nature used externally to soothe or protect.

emphysema: An abnormal swelling of body tissues caused by the accumulation of air.

endosperm: *bot.* Nutritive matter in seed plant ovules.

entire: *bot.* Having no teeth or indentations (in reference to leaves).

expectorant: A substance used to expel mucus from the respiratory tract.

febrifuge: A substance that reduces or prevents fever.

fetid: Having a bad smell; stinking.

flatulence: Gas in the stomach or intestines.

galactagogue: An agent that encourages or increases the secretion of milk.

genus: *bot.* A taxonomic term: a genus is the main subdivision of a family and includes one or more species.

glabrous: *bot.* Not hairy.
glaucous: *bot.* Covered with bloom.
globose: *bot.* Approximately spherical.
gout: A disturbance of the metabolism, characterized by an excess of uric acid in the blood and deposits of uric acid salts in the tissues around the joints.
hemostatic: An agent that arrests bleeding and hemorrhages.
hepatic: A compound that acts on the liver.
herb: *bot.* A plant that has no woody tissue and that dies down to the ground at the end of a growing season.
hoary: *bot.* Closely covered with short and fine whitish hairs.
incised: *bot.* Marked or scarred.
infusion: A preparation made by steeping plant parts in hot water.
irritant: A substance that produces irritation or inflammation of the skin or internal tissue.
jaundice: A disease in which the body becomes abnormally yellow, caused by the presence of bile pigments in the blood.
lanceolate: *bot.* Widening to a maximum near the base and tapering to a point at the apex.
laxative: An agent promoting evacuation of the bowels or a mild purgative.
leukorrhea: A morbid, whitish discharge from the vagina and uterus, usually resulting from chronic infection.
liniment: A medicinal substance, thinner than an ointment, that is gently rubbed into the skin for sprains and bruises.
maceration: An herbal preparation made by chewing, mashing, crushing, or blending to extract the juices from the plant's tissues.
migraine: A severe type of periodically returning headache.
morphine: A derivative of opium used to induce sleep and relieve pain.

mucilaginous: Characterized by a gummy or gelatinous consistency.
mucus: The thick, slimy secretion of the mucous membranes.
nervine: An agent that has a calming or soothing effect on the nerves.
neuralgia: A severe recurrent pain along one or more nerves, usually not associated with changes in the nerve structure.
obtuse: *bot.* Rounded or blunt.
opposite: *bot.* Growing two to a node on opposite sides.
ovate: *bot.* Oval-like shape.
panicle: *bot.* A raceme compounded by branching.
pectoral: A remedy for pulmonary or other chest diseases.
perennial: *bot.* Living through three or more seasons.
pinnate: *bot.* Having leaflets arranged in opposite rows along the petiole (leaf stalk).
poultice: A mixture of various healing herbs made into a paste wrapped in material (e.g., cheese cloth), and applied to the surface of the skin.
procumbent: *bot.* A plant or stem laying on the ground without making roots.
pubescent: *bot.* Covered with down or soft, short hairs.
purgative: An agent that produces a vigorous emptying of the bowels.
pyorrhea: An infection of the gums and tooth sockets, characterized by the formation of pus.
raceme: *bot.* A variety of flower clusters in which single flowers grow individually on small stems arranged at intervals along a single larger stem.
rhizome: *bot.* An underground portion of a stem, producing shoots on top and roots beneath; different from a root in that it has buds, nodes, and scaly leaves; same as rootstock.

saponin: *bot.* Amorphous glucosidal steroid compounds in plants that have soap-like properties.

sedative: A soothing agent that reduces nervousness, distress, and irritation.

serrate: *bot.* Saw-toothed, with the teeth pointing toward the apex.

sessile: *bot.* Having no stalk.

spleen: A large, vascular, ductless organ in the upper left part of the abdominal cavity near the stomach.

spore: *bot.* A one-celled reproductive body produced by relatively primitive plants.

stimulant: An agent that stimulates the body or a particular organ or tissues.

stomachic: An agent that strengthens, stimulates, or tones the stomach.

sudorific: An agent that promotes or increases perspiration.

tannin: A yellowish astringent acid.

tincture: An herbal solution preserved in alcohol (see "Making Herbal Preparations").

tonic: An agent taken over a period of time—in frequent, small dosages—that tends to restore health and stimulate the organism.

umbel: *bot.* A more or less flat-topped flower cluster in which the pedicels (rays) arise from a common point. In compound umbels, each primary ray terminates in a secondary umbel.

vermifuge: An agent that causes the expulsion of intestinal worms.

vulnerary: A healing application for wounds.

whorl: *bot.* A circular arrangement of three or more leaves, flowers, or other pairs at the same point or level.

Bibliography

Aloe Vera Studies Organization. "13 Ways Aloe Can Help You." *Alternative Medicine*, March, 1999.
Arvigo, Rosita, and Michael Balick. *Rainforest Remedies*, Twin Lakes, WI: Lotus Press, 1993.
Barrantes, Uriel. *Huertos mixtos tropicales: Características y ventajas*. San José, Costa Rica: ITCR, 1987.
Castro, Juan José Retana. *La Huerta*, Boletín #10, MAG, San José, Costa Rica, 1992.
Gage, Diane. *Aloe Vera: Nature's Soothing Healer*. Rochester, VT: Healing Arts Press, 1988.
Gomez, Luis D. *Vegetación de Costa Rica*. San José, Costa Rica: Editorial UNED, 1986.
Etkin, Nina and Heather McMillen. "The Ethnobotany of Noni (Morina citrifolia L., Rebiaceae)." Univ. Hawaii, 2003.
Herrera, Wilberth. *Clima de Costa Rica*. San José, Costa Rica: Editorial UNED, 1986.
Holdridge, L.E. and L.J. Poveda A. *Arboles de Costa Rica*. San José, Costa Rica: CCT, 1975.
Jones, Kenneth. *Pau d'Arco: Immune Power from the Rainforest*. Rochester, VT: Healing Arts Press, 1993.
Kadans, Joseph M. *Encyclopedia of Fruits, Vegetables, Nuts and Seeds for Healthful Living*. West Nyack, NY: Parker Publishing,1973.
Lust, John. *The Herb Book*. New York: Bantam, 1974.
Majeed, Muhammed, Vladimir Badmaev and Frank Murray. *Turmeric and the Healing Curcuminoids*. New Cannan, CT: Keats Publishing, Inc., 1996.
Meltzer, Sol. *Herb Gardening in the South*. Houston: Pacesetter Press, 1977.
Mollison, Bill and David Holmgren. *Permaculture One and Two*. Winters, CA: Tagari Press, 1981.
Morton, Julia F. *Fruits of Warm Climates*. Florida Flair Books, 1987.
Murray, Michael T. *The Healing Power of Herbs: The Enlightened Person's Guide to the Wonders of Medicinal Plants*. Rocklin, CA: Prima Publishing, 1991.
Nirav, Shunyam. *Hawaiian Organic Growing Guide*. Maui: Oasis Maui, 1987.
Núñez Meléndez, Esteban. *Plantas medicinales de Costa Rica y su folclore*. San José, Costa Rica: Editorial Universidad de Costa Rica, 1975.
Ocampo, Rafael A. *Jardines para la salud*. San José, Costa Rica: ITCR, 1986.
Pittier, Henri. *Plantas usuales de Costa Rica*. San José, Costa Rica: Editorial Costa Rica, 1978.
Robineau, L. (ed.), *TRAMIL Farmacopea Vegetal Caribeña*. Nicaragua: Editorial Universitaria, UNAN-León, 2005.
Rodale, Robert, *How to Grow Vegetables and Fruits bv the Organic Method*. Emmaus, PA: Rodale Press, 1977.
Sarkis, Alia and Victor Campos. *Curanderismo curanderismo tradicional del costarricense*. San José, Costa Rica: Editorial de Costa Rica, 1981.
Schulick, Paul. *Ginger: Common Spice and Wonder Drug*. Brattleboro, VT: Herbal Free Press, 1993.
Schultes, Richard and Robert Raffauf. *The Healing Forest*. Portland: Dioscorides Press, 1995.
Szekely, Edmund Bordeaux. *The Ecological Health Garden*. Cartago, Costa Rica: IBS, 1977.

Vargas Chinchilla, Seidy. *Plantas medicinales: la naturaleza como guardián de su salud.* San José, Costa Rica: EEDCAS, 1990.

Westbrooks, R.G. and J.W. Preacher. *Poisonous plants of Eastern North America.* Columbia, South Carolina: University of South Carolina Press, 1986.

White, Alan. *Hierbas del Ecuador.* Quito, Ecuador: Ediciones Libri Mundi, 1976.

Whitaker, Dr. Julian. "A Natural Sweetener That's Also Calorie Free." *Health & Healing,* Vol. 4, No. 12, December, 1994.

Visual Index

Pimenta dioica • 9
Allspice, Jamaica

Aloe vera • 10
Aloe vera, Sábila

Bixa orellana • 12
Annatto, Achiote

Chaptalia nutans • 13
Arnica, Árnica falsa

Artemisia spp. • 14
Artemisia, Ajenjo

Persea americana • 15
Avocado, Aguacate

Ocimum basilicum L. • 16
Basil, Albahaca

Momordica charantia • 17
Bitter gourd, Pepinillo or sorosí

Quassia amara • 18
Bitterwood, Hombre grande

Justicia tinctoria • 19
Bluebush, Azul de mata

Borago officinalis • 20
Borage, Borraja

Sida rhombifolia L. • 21
Broom weed, Escobilla

Justicia pectoralis • 22
Carpenter's bush, Tilo or tila

Ricinus communis • 23
Castor bean, Higuerilla

Capsicum frutesens • 24
Cayenne, Chile picante

Medicinal Plants of Costa Rica / **131**

Cecropia spp. • 25
Cecropia tree, Guarumo

Matricaria chamomilla and *Anthemis noblilis* • 27
Chamomile, Manzanilla

Hyptis suaveolens • 28
Chan

Smilax lanceolata • 29
China root, Cuculmeca

Coriandrum sativum • 30
Cilantro, Culantro

Cinnamomum verum • 31
Cinnamon, Canela

Citratus spp. • 32
Citrus, Cítrico

Cocos nucifera • 34
Coconut, Coco

Zea mays • 35
Cornsilk, Pelo de maíz

Taraxacum officinale • 36
Dandelion, Diente de león

Caesalpinia pulcherrima • 37
Dwarf poinciana, Hoja sen

Echinacea purpuria and *E. longata* • 38
Echinacea

Sambucus mexicana • **39**
Elderberry, Saúco

Eucalyptus cinerea and other species • 41
Eucalyptus, Eucalypto

Tanacentum parthenium • 42
Feverfew, Altamisa

Allium sativum • 43
Garlic, Ajo

Zingiber officinale • 44
Ginger, Jengibre

Cassia fistula • 46
Golden shower tree, Cañafístula

132 / Medicinal Plants of Costa Rica

Centella asiatica (formerly *Hydrocotyle asiatica*) • 47
Gotu kola

Plantago major • 48
Greater plantain, Llantén

Hymenaea courbaril • 49
Guapinol

Psidium guajava • 51
Guava, Guayaba

Bursera simaruba • 52
Gumbo limbo, Indio desnudo or jiñocuave

Hibiscus spp. • 53
Hibiscus, Amapola

Piper auritum • 55
Hoja de estrella

Equisetum spp. • 56
Horsetail, Cola de caballo

Terminalia catappa • 57
Indian almond, Almendro

Neuroloena lobata • 59
Jackass bitters, Gavilana

Lippia alba • 60
Juanilama

Cymbopogon citratus • 61
Lemongrass, Zacate de limón

Kalanchoe pinnata • 62
Life everlasting, Hoja del aire

Mimosa pudica • 63
Mimosa, Dormilona

Satureja viminea • 64
Mint, Menta de palo

Triumfetta semitriloba • 66
Mozote de caballo

Morinda citrifolia • 67
Noni

Lippia graveolens and *Origanum vulgare* • 69
Oregano, Orégano

Medicinal Plants of Costa Rica / **133**

Carica papaya • 70
Papaya

Petroselinium crispum • 71
Parsley, Perejil

Passiflora quadrangularis • 72
Passion flower, Granadilla real

Mentha spp. • 73
Peppermint and Spearmint, Hierbabuena

Cassia grandis • 75
Pink shower tree, Carao

Tabebuia rosea, T. impetiginosa • 76
Pink trumpet tree and Pau d'arco, Roble de sabana and Pau d'arco

Senna occidentalis • 78
Pisabed, Pico de pájaro

Phytolacca decandra and *P. rugosa* • 80
Pokeweed, Jaboncillo

Opuntia tuna and *Nopalea cochenillifera* • 81
Prickly pear cactus, Tuna

Spondias purpurea • 82
Purple mombin, Jocote

Gliricidia sepium • 83
Rat killer tree, Madero negro

Croton gossypifolius • 84
Red targua, Targuá colorado

Hamelia patens • 85
Redhead, Zorrillo real

Catharanthus roseus • 87
Rose periwinkle, Mariposilla

Rosmarinus officinalis • 88
Rosemary, Romero

Ruta graveolens • 89
Rue, Ruda

Salvia officinales • 90
Sage, Salvia

Sansevieria trifasciata • 91
Sansevieria, Lengua de suegra

134 / Medicinal Plants of Costa Rica

Senna reticulata • 92
Saragundí

Smilax medica, S. regelii, and *S. ornata* • 93
Sarsaparilla, Zarzaparrilla

Heliotropium peruvianum • 94
Scorpion's tail, Cola de alacrán

Lantana camara • 95
Shrub verbena, Lantana

Costus spicatus and *C. ruber* • 96
Spiral flag, Caña agria

Stevia rebaudiana • 98
Stevia, Estevia, hierba dulce, or hoja dulce

Urtica dioica • 99
Stinging nettle, Ortiga

Ficus glabrata and *F. jimenezii* • 100
Strangler fig, Higuerón

Thymus vulgaris • 101
Thyme, Tomillo

Tradescantia zebrina • 102
Tradescantia, Hoja del milagro

Cedrela mexicana • 103
Tropical cedar, Cedro

Guazuma ulmifolia • 104
Tropical elm, Guácimo

Vernonia canescens H.B.K. • 105
Tuete

Curcuma longa • 106
Turmeric, Cúrcuma

Verbena litorales • 108
Vervain, Verbena

Buddleja americana • 109
Wild sage, Salvia virgen

Acnistus arborescens • 111
Wild tobacco tree, Güitite

Simarouba glauca • 112
Winged-leaved quassia, Aceituno negro

Medicinal Plants of Costa Rica / **135**

Chenopodium ambrosioides • 113
Wormseed, Apazote

Achillea millefolium • 114
Yarrow, Milenrama

Rumex crispus • 115
Yellow Dock, Romaza or ruibarbillo

Yucca guatemalensis • 117
Yucca, Itabo

Blechum pyramidatum • 118
Zornia, Trencilla

136 / Medicinal Plants of Costa Rica

Index

Aceituno negro, 112
Achillea millefollium, 114
Achiote, 12
Acnistus arborescens, 111
Aguacate, 15
Ajenjo, 14
Ajo, 43
Albahaca, 16
Allium sativum, 43
Allspice, 9
Almendro, 57
Aloe vera, 10, 62, 66, 70, 81, 91, 107, 121
Altamisa, 42
Amapola, 53
Annatto, 12
Anthemis noblilis, 27
Apazote, 113
Arnica, 13, 36
Árnica falsa, 13
Artemisia, 14
Artemisia spp., 14
Avocado, 15
Azul de mata, 19
Bajaro, 78
Bandera española, 95
Basil, 16
Bitter gourd, 17
Bitterwood, 18, 120
Bixa orellana, 12
Blechum pyramidatum, 118
Bluebush, 19
Borage, 20
Borago officinalis, 20
Borraja, 20
Broom weed, 21
Buddleja americana, 109
Bursera simaruba, 52
Caesalpinia pulcherrima, 37
Caña agria, 96
Cañafístula, 46

Canela, 31
Capsicum frutesens, 24
Carao, 75
Carica papaya, 70
Carpenter's bush, 22
Cassia fistula, 46
Cassia grandis, 75
Castor bean, 23
Catharanthus roseus, 87
Cayenne, 24, 32
Cecropia tree, 25
Cecropia spp., 25
Cedrela mexicana, 103
Cedro, 103
Centella asiatica, 47
Chamomile, 27
Chan, 28
Chaptalia nutans, 13
Chenopodium ambrosioides, 113
Chicory, 36
Chilamate blanco, 100
Chilamates, 100
Chile picante, 24
China root, 29
Cilantro, 30, 45
Cinco negritos, 95
Cinnamomum verum, 31
Cinnamon, 29, 31
Citratus spp., 32
Cítrico, 32
Citrus, 32, 68
Coco, 34
Coconut, 34, 96
Cocos nucifera, 34
Cola de alacrán, 94
Cola de caballo, 56
Cone flower, 38
Corallilo, 85
Coriander, 30
Coriandrum sativum, 30

Medicinal Plants of Costa Rica / **137**

Corn, 35, 83
Cornsilk, 35
Cortez negro, 76
Costus ruber, 96
Costus spicatus, 96
Croton gossypifolius, 84
Cucaracha, 102
Cuculmeca, 29
Culantro, 30
Culantro coyote, 30
Cumin, 45
Cúrcuma, 106
Curcuma longa, 106
Cymbopogon citratus, 61
Dandelion, 13, 36
Diente de león, 36
Dormilona, 63
Dwarf poinciana, 37
Dysentery bark, 112
Echinacea, 38
Echinacea longata, 38
Echinacea purpuria, 38
Elderberry, 39
Equisetum spp., 56
Eryngium foetidum L., 30
Escobilla, 21
Estevia, 98
Eucalypto, 41
Eucalyptus, 41
Eucalyptus cinerea, 41
Feverfew, 42
Ficus glabrata, 100
Ficus jimenezii, 100
Frijolillo, 78
Frijolillo negro, 78
Garlic, 24, 32, 43
Gavilana, 59
Ginger, 33, 44, 60, 107, 120, 121
Gliricidia sepium, 83
Golden seal, 115
Golden shower tree, 46
Gotu kola, 47
Granadilla real, 72

Greater plantain, 48
Guácimo, 104
Guapinol, 49
Guarumo, 25
Guava, 51
Guayaba, 51
Guazuma ulmifolia, 104
Güitite, 111
Gumbo limbo, 52
Hamelia patens, 85
Heliotropium indicum, 94
Hibiscus, 53
Hibiscus spp., 53
Hierba dulce, 98
Hierbabuena, 73
Higuerilla, 23
Higuerón, 100
Higuerón blanco, 100
Hoja de estrella, 55
Hoja del milagro, 102
Hoja del aire, 62
Hoja dulce, 98
Hoja sen, 37
Hombre grande, 18
Horsetail, 56
Hydrastis canadensis, 115
Hydrocotyle asiatica, 47
Hymenaea courbaril, 49
Hyptis suaveolens, 28
Indian almond, 57
Indian mulberry, 67
Indio desnudo, 52
Ipe, 76
Itabo, 117
Jaboncillo, 80
Jackass bitters, 59
Jamaica, 9
Jamaican hibiscus, 33, 44, 54
Jengibre, 44
Jiñocuave, 52
Jocote, 82
Juanilama, 60
Justicia pectoralis, 22

Justicia tinctoria, 19
Kalanchoe pinnata, 62
Kī kuán, 94
Lantana, 95
Lantana camara, 95
Lapacho, 76
Lemongrass, 33, 44, 60, 61
Lengua de suegra, 91
Life everlasting, 62
Lippia alba, 60
Lippia graveolens, 69
Llantén, 48
Madero negro, 83
Manzanilla, 27
Mariposilla, 87
Matricaria chamomilla, 27
Menta de palo, 64
Mentha spp., 73
Milenrama, 114
Mimosa, 63
Mimosa pudica, 63
Mint, 44, 64, 73, 89
Mints, 30, 65, 73, 74
Momordica charantia, 17
Morinda citrifolia, 67
Mozote de caballo, 66
Mugwort, 14
Naranja agria, 33
Neuroloena lobata, 59
Noni, 67
Nopalea cochenillifera, 81
Nutmeg, 29
Ocimum basilicum L., 16
Opuntia tuna, 81
Orchids, 111
Oregano, 69, 89, 101
Origanum vulgare, 69
Ortiga, 99
Palo camarón, 85
Papaya, 11, 70
Parsley, 71
Passiflora quadrangularis, 72
Passion flower, 72

Pau d'arco, 76
Pelo de maíz, 35
Pepinillo, 17
Peppermint, 65, 73
Perejil, 71
Persea americana, 15
Petroselinium crispum, 71
Phytolacca decandra, 80
Phytolacca rugosa, 80
Pico de pájaro, 78
Pimenta dioica, 9
Pineapple, 34, 68, 82, 96
Pink shower tree, 75
Pink trumpet tree, 76
Pipa, 34
Piper auritum, 55
Pisabed, 78
Plantago major, 48
Pokeweed, 80
Prickly pear cactus, 81
Psidium guajava, 51
Purple mombin, 82
Quassia amara, 18
Rat killer tree, 83
Red targua, 84
Redhead, 85
Rhubarb, 115
Ricinus communis, 23
Roble colorado, 76
Roble de sabana, 76
Romaza, 115
Romero, 88
Rose periwinkle, 87
Roselle, 54
Rosemary, 88, 101
Rosmarinus officinalis, 88
Ruda, 89
Rue, 89
Ruibarbillo, 115
Rumex crispus, 115
Rumex rhaponticum, 115
Ruta graveolens, 89
Sábila, 10

Sage, 90
Salvia, 90
Salvia officinales, 90
Salvia virgen, 109
Sambucus mexicana, 39
Sansevieria, 91
Sansevieria trifasciata, 91
Saragundí, 92
Sarsaparilla, 29, 93
Satureja viminea, 64, 73
Saúco, 39
Scorpion's tail, 94
Senna occidentalis, 78
Senna reticulata, 92
Shrub verbena, 95
Sida rhombifolia L., 21
Simarouba glauca, 112, 118
Smilax lanceolata, 29
Smilax medica, 93
Smilax ornata, 93
Smilax regelii, 93
Sonajera, 78
Sorosí, 17
Soterrey, 95
Spearmint, 65, 73
Spiral flag, 34, 96
Spondias purpurea, 82
Stevia, 65, 74, 98
Stevia rebaudiana, 98
Stinging nettle, 99
Stinky toe, 49
Strangler fig, 100
Tabebuia impetiginosa, 76
Tabebuia pentaphylla, 76
Tabebuia rosea, 76
Taheebo, 76
Tanacentum parthenium, 42
Taraxacum officinale, 13, 36

Targuá colorado, 84
Terminalia catappa, 57
Thyme, 101
Thymus vulgaris, 101
Tila, 22
Tilo, 22
Tomillo, 101
Tradescantia, 102
Tradescantia zebrina, 102
Trencilla, 118
Triumfetta semitriloba, 66
Tropical cedar, 103
Tropical elm, 104
Tuete, 105
Tuna, 81
Turmeric, 45, 106, 120, 121
Urtica dioica, 99
Verbena, 108
Verbena litorales, 108
Vernonia canescens H.B.K., 105
Vervain, 108
Wandering jew, 102
Wild coriander, 30
Wild sage, 109
Wild tobacco tree, 111
Winged-leaved quassia, 112, 118
Wormseed, 113
Yarrow, 114
Yellow dock, 115
Yucca, 117
Yucca guatemalensis, 117
Zacate de limón, 61
Zarzaparrilla, 93
Zea mays, 35
Zingiber officinale, 44
Zornia, 118
Zorrillo real, 85